VISUAL SPATIAL INSPIRATION

TOP

SPACE & ART

IV

ARTPOWER

Top Space & Art IV

ARTPOWER™

Designer: Zheng Qin
Chief Editor: Li Aihong

Address: Room C, 9/F., Sun House, 181 Des Voeux Road Central, Hong Kong, China
Tel: 852-31840676
Fax: 852-25432396

Editorial Department
Address: G009, Floor 7th, Yimao Centre, Meiyuan Road, Luohu District, Shenzhen, China
Tel: 86-755-82913355
Fax: 86-755-82020029

URL: www.artpower.com.cn
E-mail: artpower@artpower.com.cn

ISBN 978-988-68243-1-3

Printed in China

PREFACE

Architecture can be very slow. Sometimes you would say it hardly moves at all. And yet even the best preserved of ancient temples is a totally different experience for us now than it was for its creators. It may be the same stone, the same wood, even the same rituals and the same smells, but we have changed – literature, cinema, music and a thousand of images have gone through our mind and changed the way we experience such apparently immovable spaces. Art has changed that space, opening the doors of perception.

Most buildings do change physically over time, with daily use, frequent adaptations, regular renovations, and eventually full upheavals. Some of these changes are almost involuntary or accidental, brought about by simple friction; but other transformations require a cultural shift, a different perception, a new understanding of the space within its stubbornly rigid frame. This is a difficult task, but Art has the power to redescribe the existing, to transform it into a new cultural environment.

With new construction, architecture should be able to move a bit faster, providing the new kinds of spaces required for the exciting new experiences we now want to live, with new technology and new economies, new social organizations, or new kinds of families and different personal situations. But the not-yet-existing is not free of inertia. The battle between the new needs and the old ways is a hard one, as we find it very difficult to visualize the experience of a new kind of space, or even the need of such novelty.

I am not talking about 3D renders; what I mean that we sometimes find it difficult to evaluate the unknown possibilities that lie beyond our comfort zone. If we want to find a new experience of space that may satisfy the new needs and be desirable by all the people involved, from designers to final users, we need to undertake the explorative journey that we generally call Art.

Art is the most powerful tool we have if we want to engage our collective imaginations into creating, understanding and finally encouraging new experiences of space. This book has plenty of examples of this; some explore the relationship between the past and the future, or between the virtual world of visual culture and the tough impenetrability of solid matter; some explore the possibilities of new technologies, be it in the production or in the display; some are charming, delicate or educated; some are brash, funky or unstable; but all of them are very seriously engaged in playful exploration.

by Carlos Arroyo

Contents

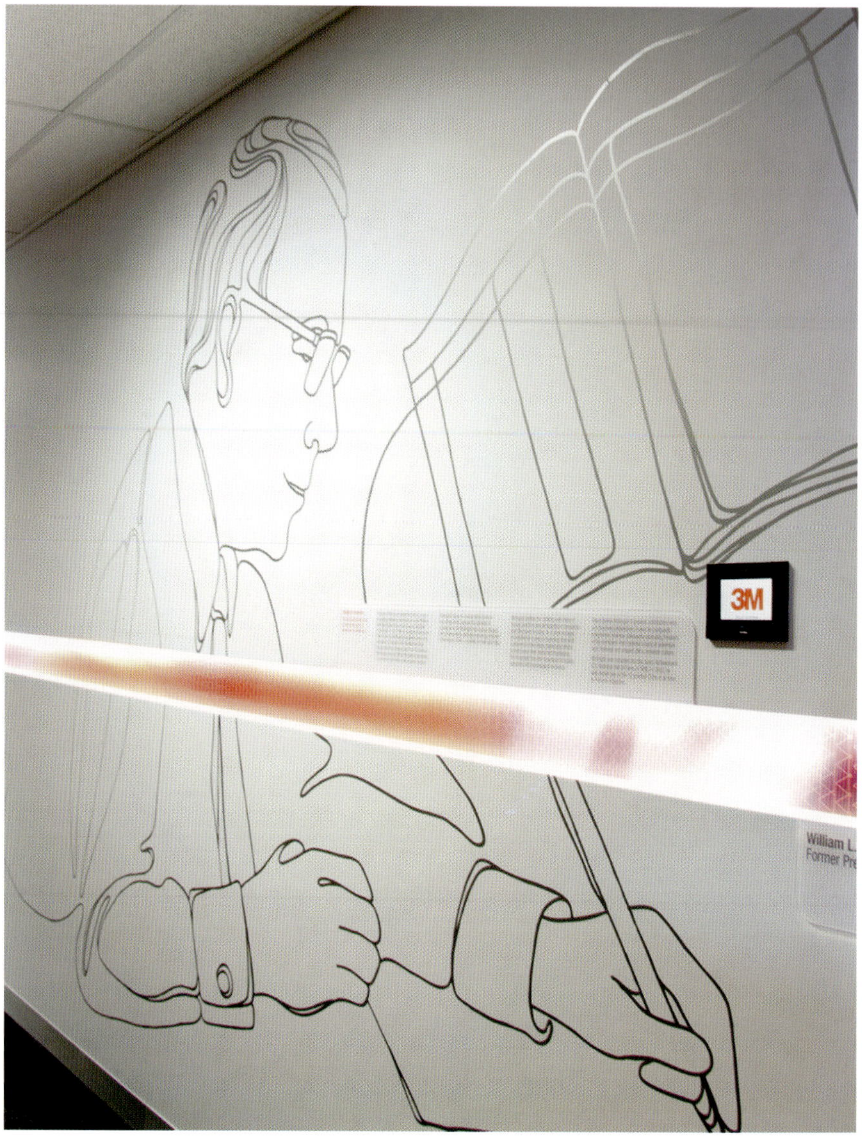

3M AUSTRALIA HQ

The designers have just completed an extensive branded environment, signage and wayfinding scheme for a truly iconic brand, 3M.

Working with such a massive brand has both its advantages and challenges and the team at THERE more than rose to the occasion. 3M's collaborative work ethos fit perfectly with their own approach that places openness, inclusiveness and transparency at the core of how we work.

Working in close partnership with the Colliers interior design team, the designers' starting point was conveying 3M's brand essence "Harnessing the chain reaction of new ideas", they developed an immersive graphic language based on geometric shapes. Each of the six levels was designated a different shape, allowing a chain reaction of subtle variation from level to emerge. This visual language flexed and adapted, it was deployed as a treatment across super scale images of 3M's notable inventors, it allowed us to tell stories, share the company's historical information and get to the heart of product insights in a cohesive and engaging way. It allowed us to bring to life the innovation and creativity 3M puts into the science behind every one of their products.

The outcome for 3M was a workplace environment that showcased their history, culture, technology and products in unexpected and delightful ways. Staff and visitors come away from the environment with a newfound, or renewed, curiosity for not just their products, but the stories behind them as well. It's a truly branded environment, purely because every element on every wall was inspired by insights, truths and stories that come from inside 3M, and that's something of which we are really proud.

MANN
ROOM

Graeme Mann
Innovator
Tamper-Evident Adhesive for Passpo

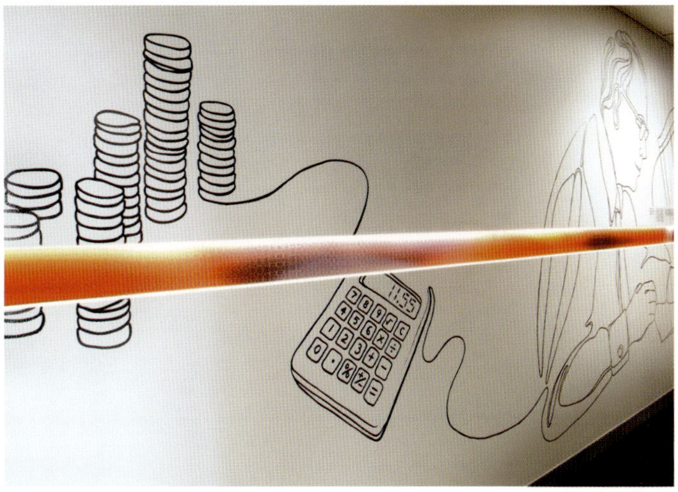

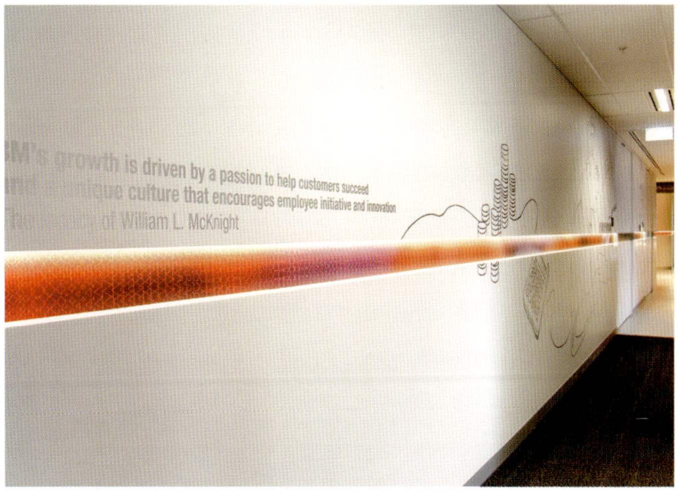
growth is driven by a passion to help customers succeed
culture that encourages employee initiative and innovation
of William L. McKnight

Design Agency
THERE Design

Client
3M

Location
Australia

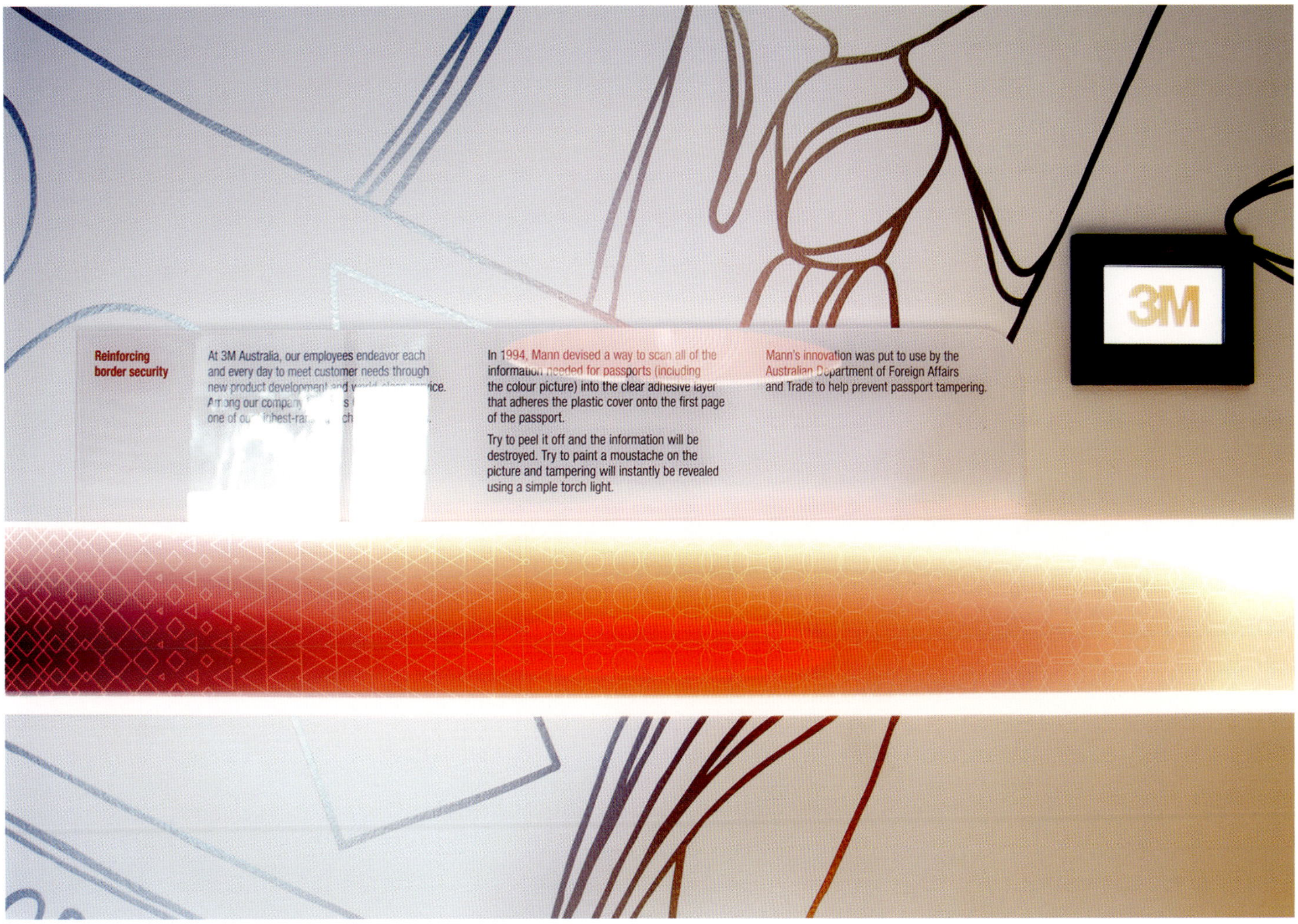

Reinforcing border security
At 3M Australia, our employees endeavor each
and every day to meet customer needs through
In 1994, Mann devised a way to scan all of the information needed for passports (including the colour picture) into the clear adhesive layer that adheres the plastic cover onto the first page of the passport.
Try to peel it off and the information will be destroyed. Try to paint a moustache on the picture and tampering will instantly be revealed using a simple torch light.
Mann's innovation was put to use by the Australian Department of Foreign Affairs and Trade to help prevent passport tampering.
3M

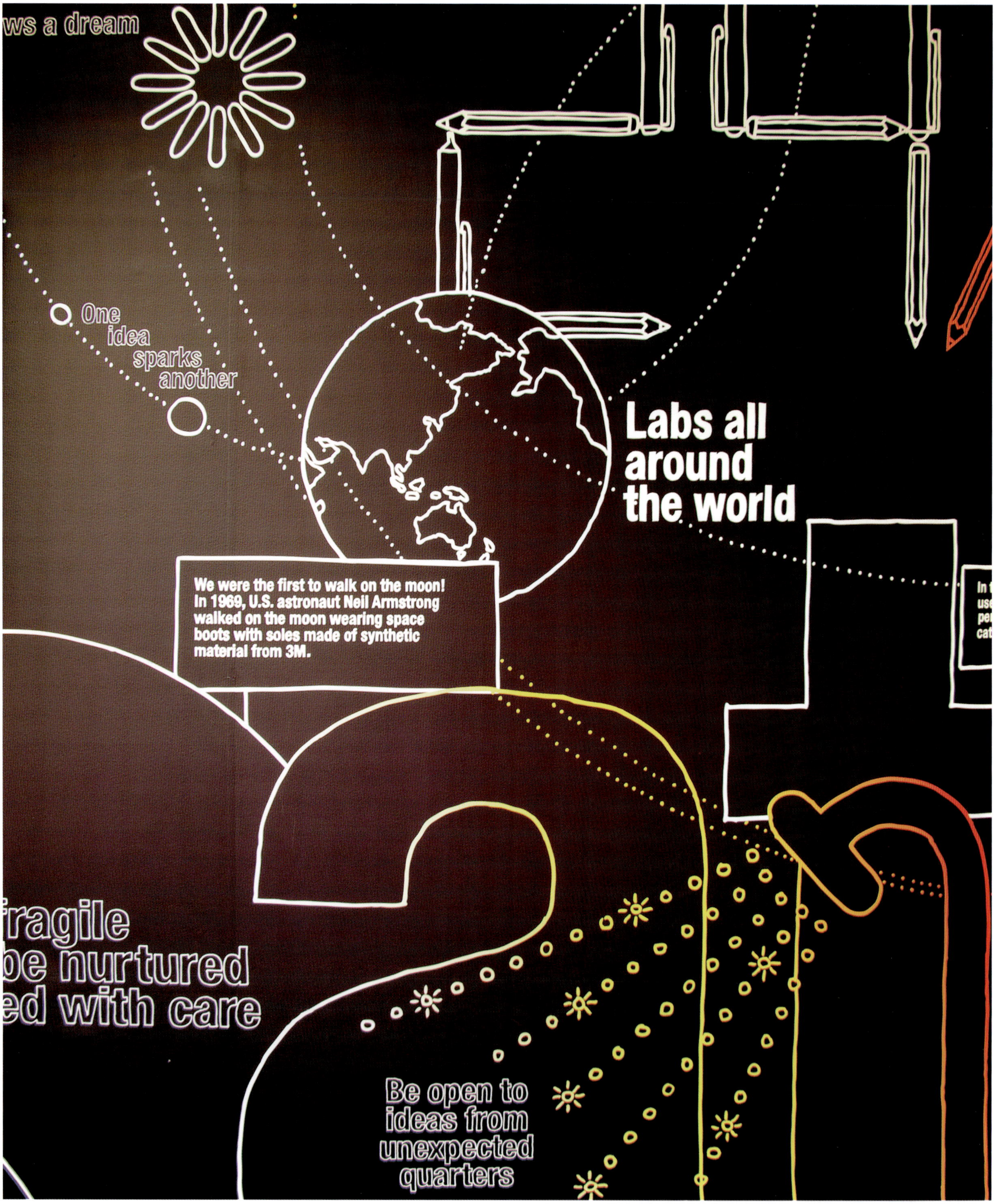
ws a dream
One idea sparks another
Labs all around the world
We were the first to walk on the moon! In 1969, U.S. astronaut Neil Armstrong walked on the moon wearing space boots with soles made of synthetic material from 3M.
fragile
be nurtured
ed with care
Be open to ideas from unexpected quarters

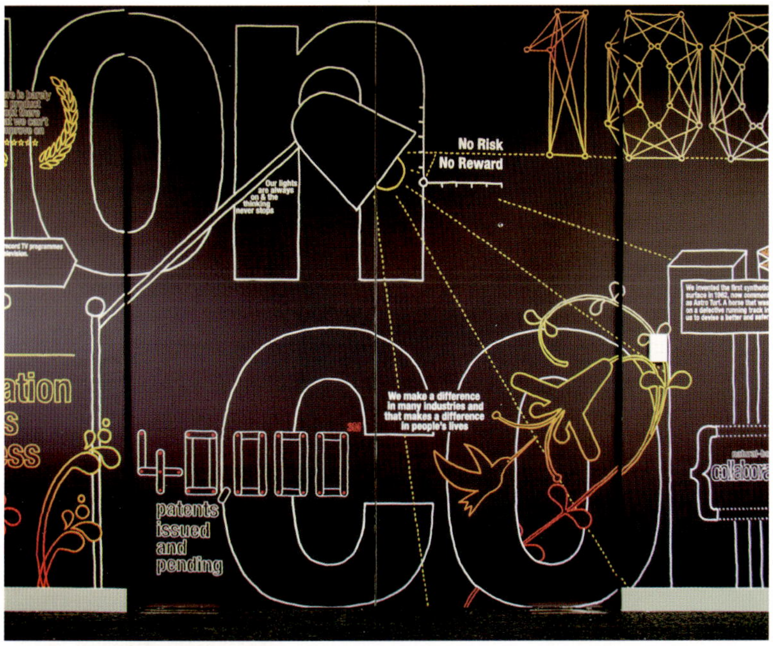
No Risk
No Reward
We make a difference in many industries and that makes a difference in people's lives
patents issued and pending

never let that standard slip
VHB tape is used to hold trains and buildings together.
What's next
3M's most recent invention, autostereoscopic enhancement film, achieves eye-popping 3-D images without the need for glasses.
We want to be the first to make our own products obsolete

3Mers are masters of harnessing the power of human collaboration
Original thinking does not come from mandates
We reward people for sharing ideas
We solve problems on a global scale
During the 1940's, we invented the Sound Recording Tape, spurred by singer Bing Crosby's interest and promotion. As a result it revolutionised the entertainment industry.
Creativity comes from freedom

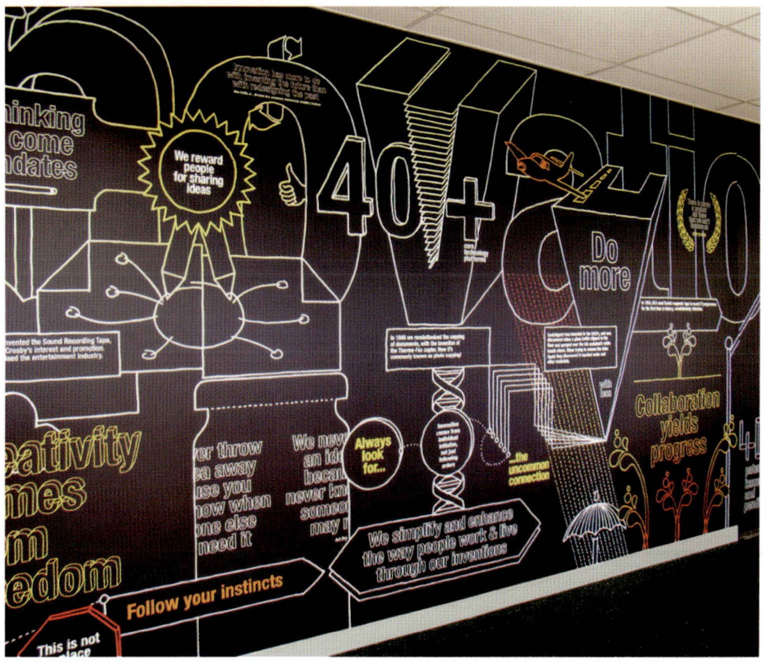
We reward people for sharing ideas
40+
Do more
Collaboration yields progress
Always look for...
...the uncommon connection
We simplify and enhance the way people work & live through our inventions
Follow your instincts

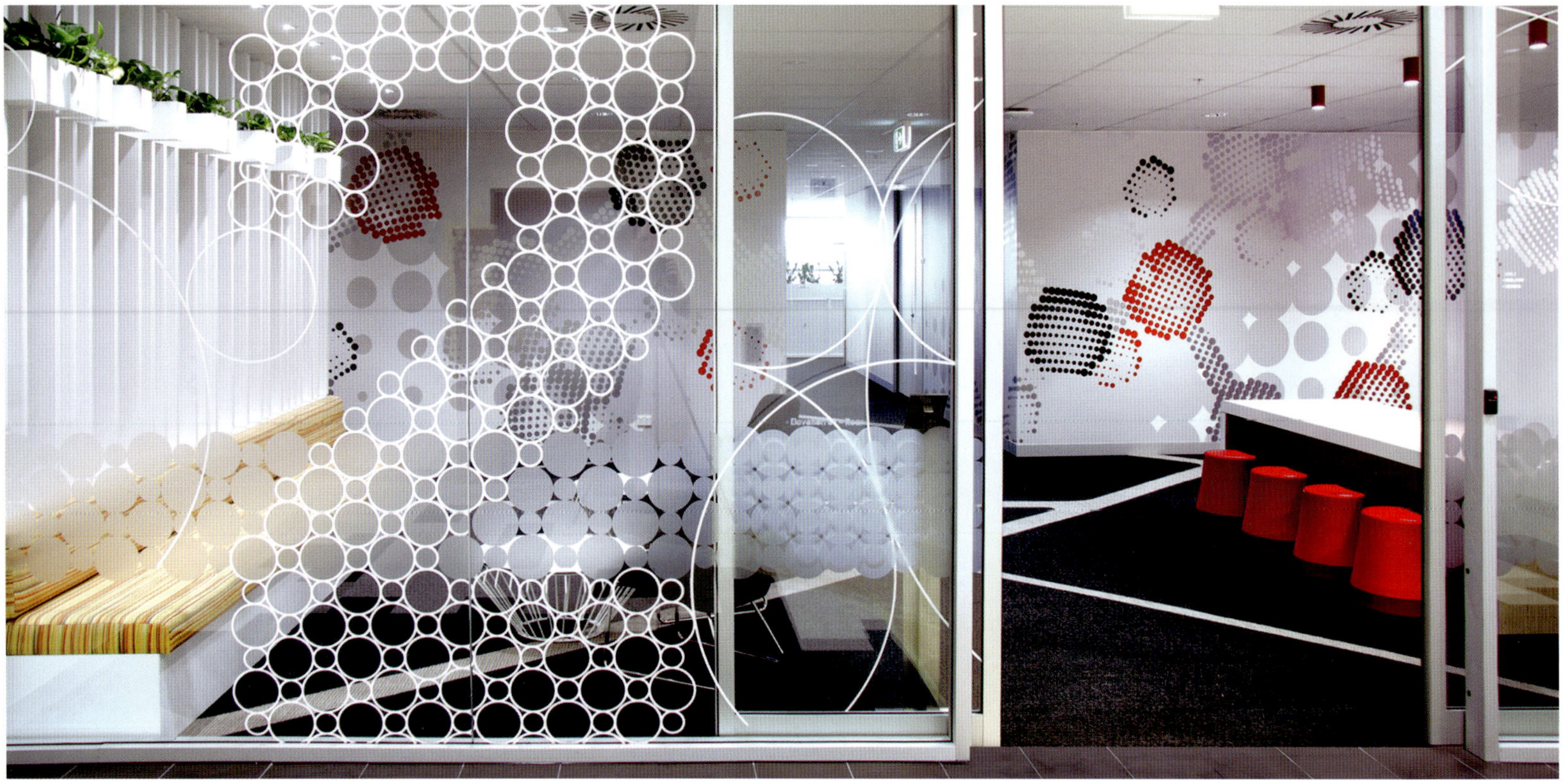

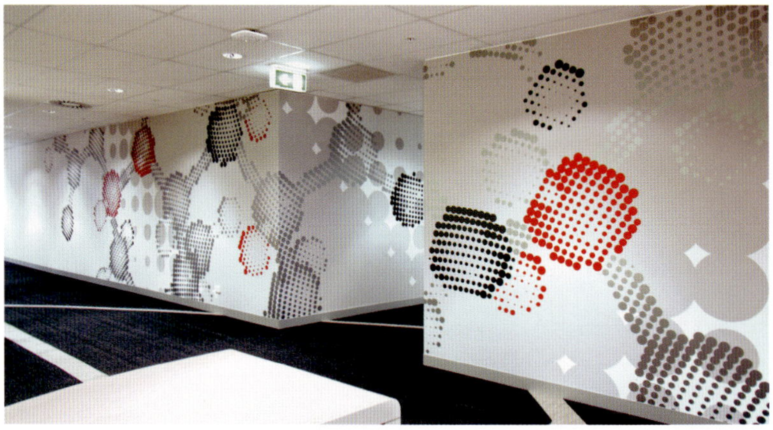

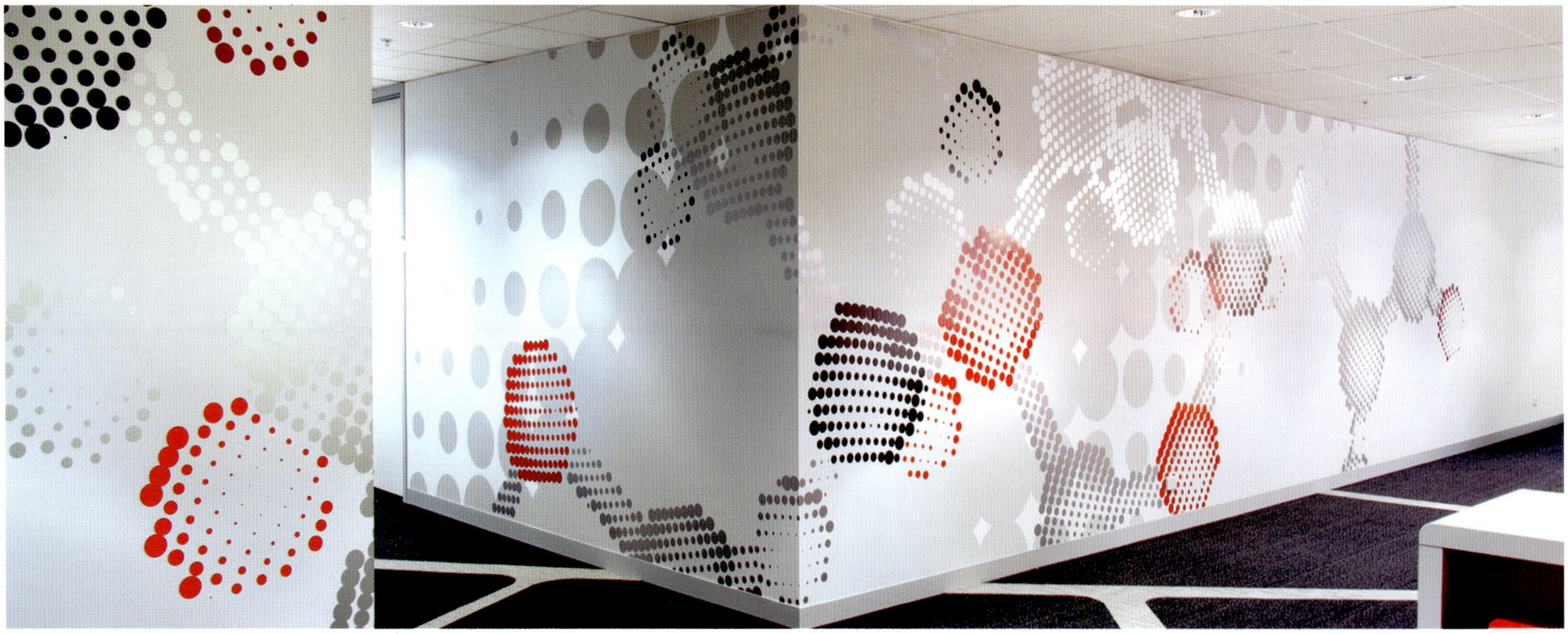

LEVEL 3
▸ ACCOUNTS PAYABLE
▸ ACCOUNTS RECEIVABLE
▸ CUSTOMER SERVICE
▸ ENGINEERING & MANUFACTURING
▸ INFORMATION TECHNOLOGY
▸ PROCUREMENT

DOUANE DOE SPACE

The douane doe space is the area of the Dutch Customs and Tax Museum focusing on children and education.

Everything in this space explanes in a comprehensive and playful manner the phenomena of borders and customs to kids. The space is devised three worlds: land, sea and sky with one overall theme: borders. A few of the elements found in the space are: a truck with it's loading space with integrated video games, a customs boat with a simulation game, an interactive scanning device to check luggage as found at the airport and last but not least the icon of the museum: Doerak the customs sniffing dog.

Design Team
Frank Tjepkema, Yannic Alidarso, Leonie Janssen

Client
Belastingdienst, Rotterdam

Location
Dutch Tax museum, Rotterdam, Holand

Area
60 m^2

Builder
Gielissen Interior & Exhibitions

Photographer
Tjep: Yannic Alidarso and Martynika Bielawska

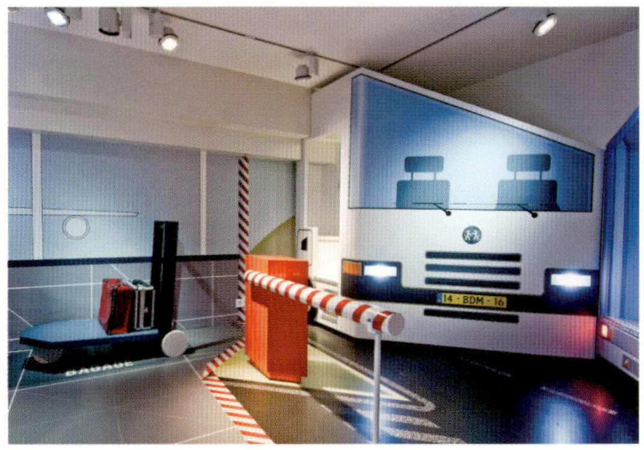

MILKYWAVE

Milkywave is the designers' installation entry for Beijing Design Week 2012. A constellation of 1664 recycled yoghurt jars re-constituted into a moebius wave of light. It hangs from the stairwell of a former bicycle factory in the historic district of Dashilar in Beijing.

For this installation the designers depart with a found object, a quintessential part of the hutong's DNA — the old yoghurt ceramic jar. Native to the Beijing cityscape millions of bottles circulate the streets uninterruptedly meeting Beijingers on their corner shop, subway station or local alley.

The designers are drawn to the pot by its materiality; translucency, color and unique shape — properties with the potential of becoming something more, and so they speculate with new forms of aggregation and clustering.

Repurposing is not only about recycling but about challenging perceptions and enabling new sensations; in milkywave they were interested in triggering memories in the visitor and freeing the jar from its original function — a way of celebrating the use of mass produced objects and promoting them as core components in the design of new systems and configurations.

In the big picture milkywave is part on an ongoing research project focused on reading and re-thinking the city through the objects and situations that inhabit it; an attempt to unveil the narratives that speak of claims and conflict, not apparent at plain sight but at the core of the dynamics that drive change in the city.

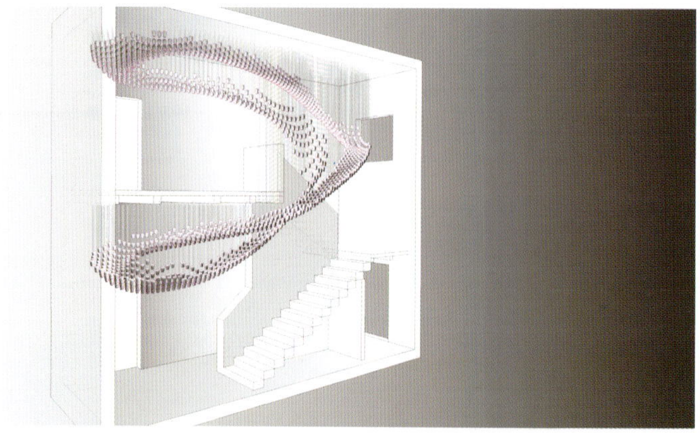

Design Agency
AIDIASTUDIO

Design Team
Rolando Rodriguez-Leal, Natalia Wrzask

Collaborator
He You

Location
Beijing, China

Photographer
AIDIASTUDIO

THE BETA MOVEMENT

Prominently sited on the Hollywood Walk-of-Fame, the Beta Movement acts as a grand-scale temporary photomaton, inviting tourists and Angelinos inside for their own red carpet moment. Stars are plucked from the famous sidewalk and projected through the gallery, producing a series of spatial distortions that are replete with filmic references to superheroes, hyperspace, hypnotism, evil lairs and astrophysical singularities. Upon entry into the space, the rough qualities of the exposed construction reveal the temporary nature of the illusion. Cutting from one void to the next, visitors make their way to a projection room where video and text describe the formal procedures at play.

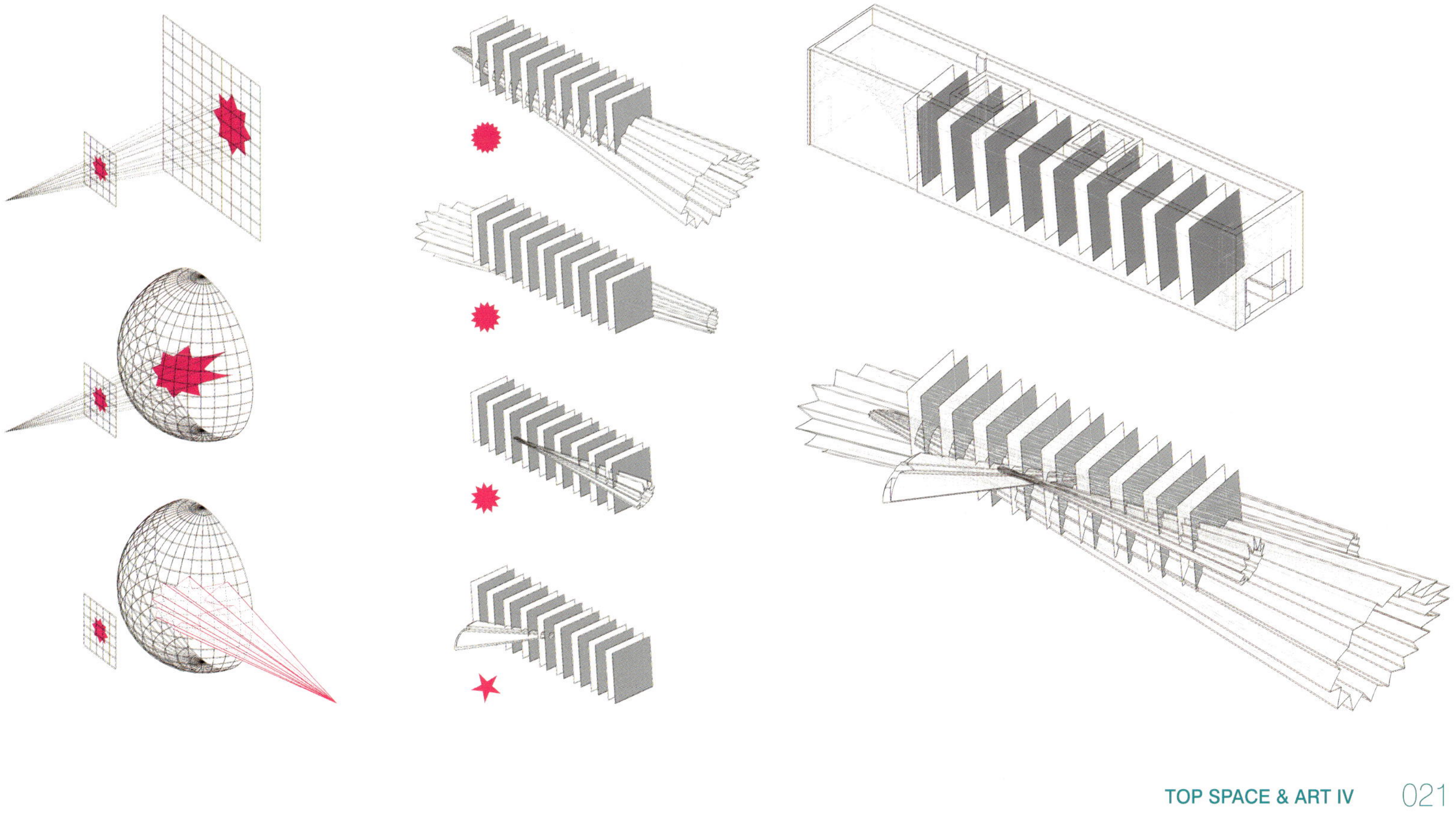

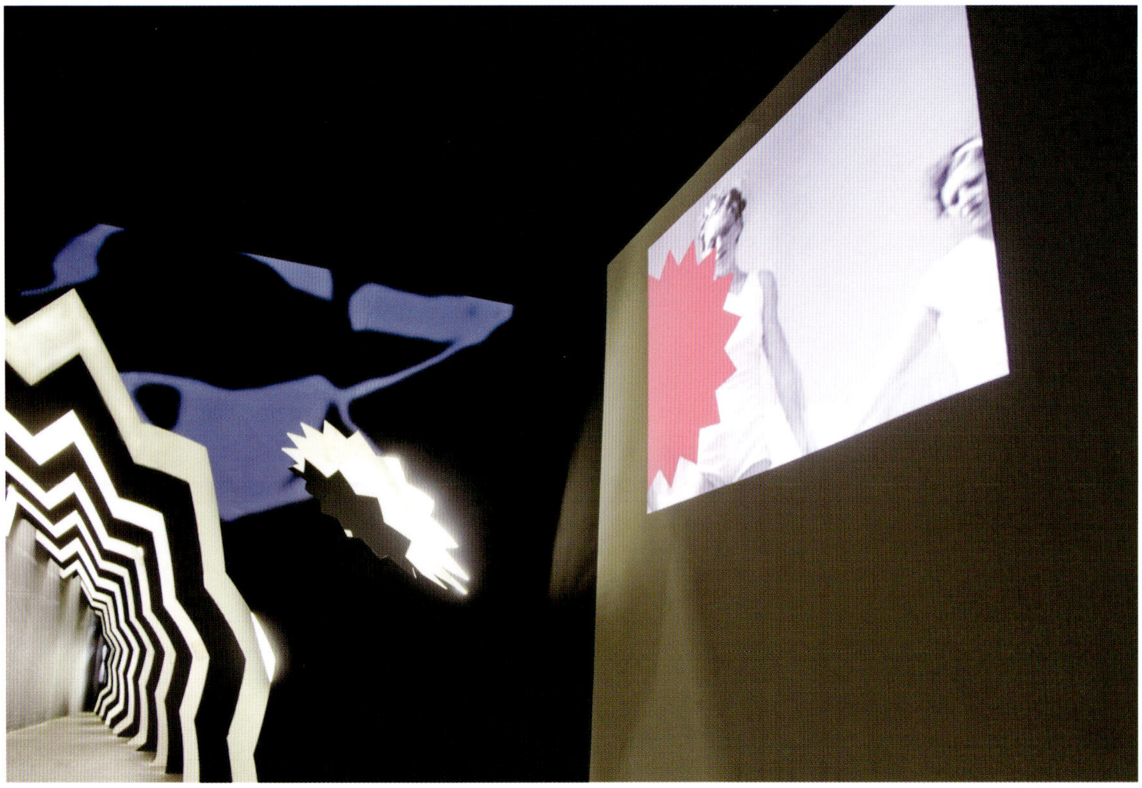

Design Agency
Spatial Ops

Team
Steven Christensen, Jean Louis Farges, Anya Sirota

Location
California, USA

Photographer
Spatial Ops

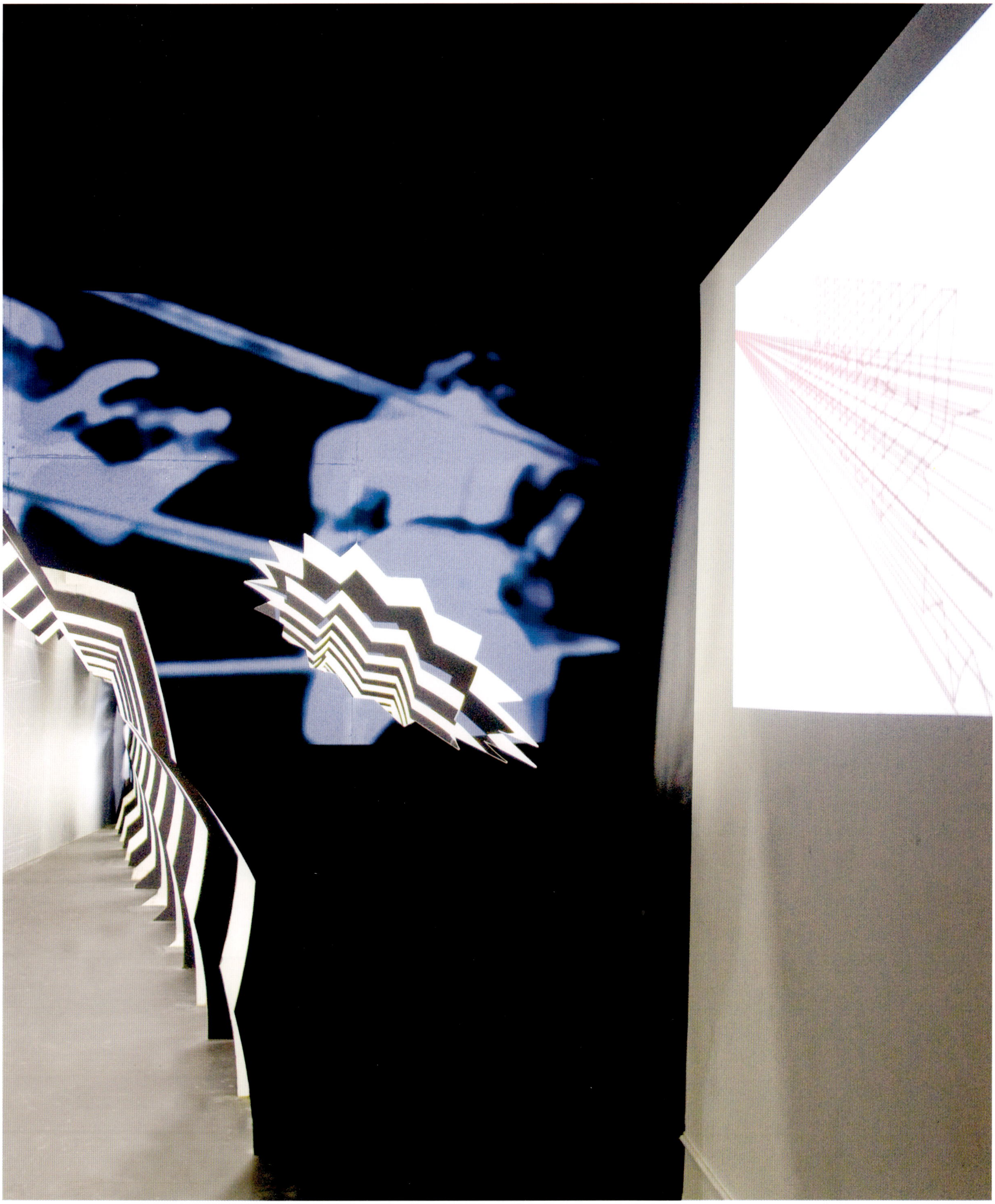

Design Agency
Synthesis Design + Architecture

Location
London, England

POLYFOLD PARTITION

The design of this dynamic wall partition was focused on the challenge of creating a three dimensional freestanding wall partition made entirely from flat sheet material. Composed entirely of a network of stacked 3D polygons, the density of the partition articulation responds to the curvature of the surface, whereas the orientation and aperture size are driven by specific view vectors from within the gallery space. Key views are highlighted, while other views are blocked. The entire assembly is made from 3 mm thick polished aluminium sheets, laser cut and etched for folding and assembly notation, and hand-folded and assembled.

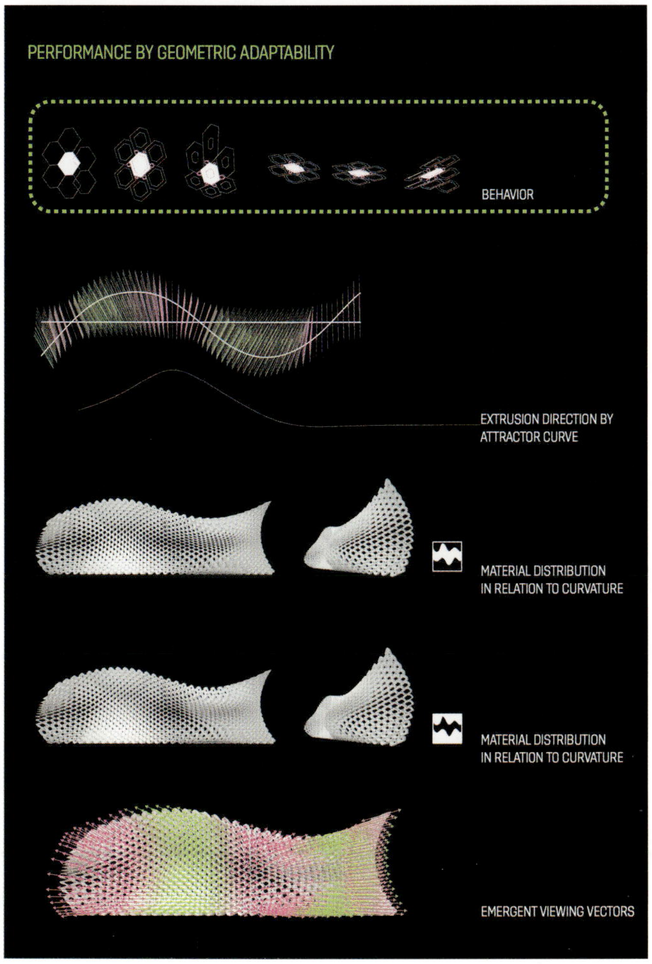
PERFORMANCE BY GEOMETRIC ADAPTABILITY
BEHAVIOR
EXTRUSION DIRECTION BY ATTRACTOR CURVE
MATERIAL DISTRIBUTION IN RELATION TO CURVATURE
MATERIAL DISTRIBUTION IN RELATION TO CURVATURE
EMERGENT VIEWING VECTORS

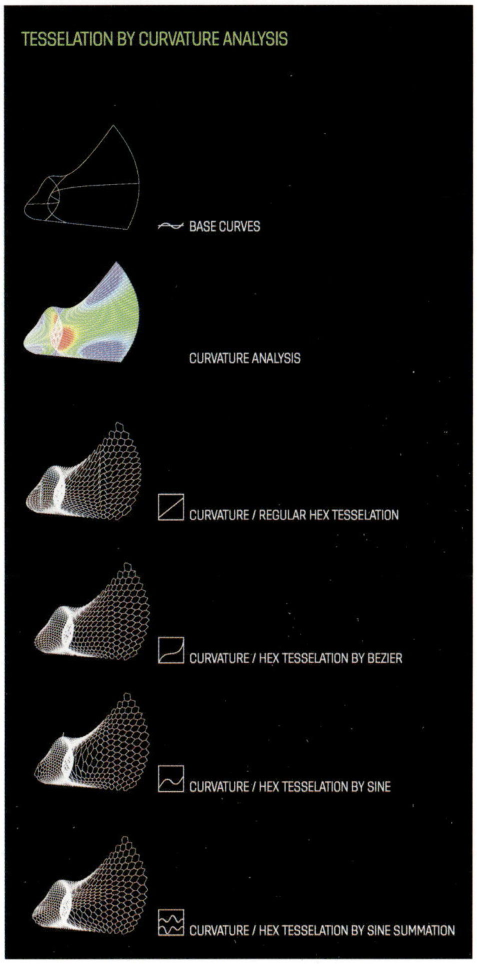
TESSELATION BY CURVATURE ANALYSIS
BASE CURVES
CURVATURE ANALYSIS
CURVATURE / REGULAR HEX TESSELATION
CURVATURE / HEX TESSELATION BY BEZIER
CURVATURE / HEX TESSELATION BY SINE
CURVATURE / HEX TESSELATION BY SINE SUMMATION

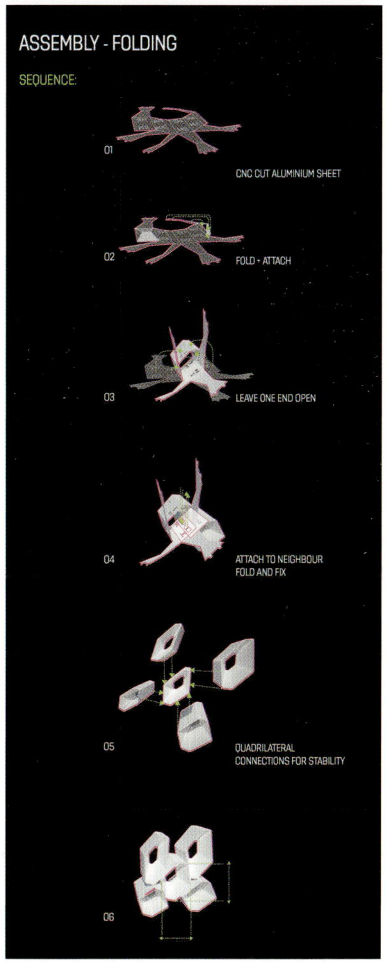
ASSEMBLY - FOLDING
SEQUENCE:
01
CNC CUT ALUMINIUM SHEET
02
FOLD + ATTACH
03
LEAVE ONE END OPEN
04
ATTACH TO NEIGHBOUR FOLD AND FIX
05
QUADRILATERAL CONNECTIONS FOR STABILITY
06

CLARIANT CORPORATE CENTRE

Clariant, an international chemical production company, commissioned Charlotte Mann to create two pieces for their corporate centre in Basel, Switzerland. The first commission was a mural that ran the length of the entrance hallway. In this mural, Mann depicts two related environments: the first is a laboratory with an experiment set up in a fume extraction cabinet and the second is the floor of a chemical plant where the same chemical process is installed on an industrial scale. The drawing deals with the physical, technical identity of the objects and treats them without hierarchy, from the postcards decorating the walls of the lab to the flanges on the pipes leaving the reactor in the chemical plant. Every element of the drawing is taken from the Clariant sites around Prateln. Mann said "the process of researching the Clariant commission was wonderful because I was given full access to a truly fascinating environment. I loved the fact that it contained objects that were almost unimaginably (to most people) hi-tech, alongside others that were utterly basic and must have been amongst the first human tools, such as a broom made of a bunch of twigs, and that both were in current use."

The second commission was to create an external piece in the forecourt of the corporate centre. Mann's suggestion was to create a wall that was a life-sized drawing of a hedgerow, as an inverse counterpart to the world of technology depicted inside, and to render the subject with the same obsessive attention to specificity and detail.

Artist
Charlotte Mann

Client
Clariant International

Location
Prateln, Basel, Switzerland

Architect
Wirth + Wirth Architekten

Photographer
Wirth + Wirth Architekten

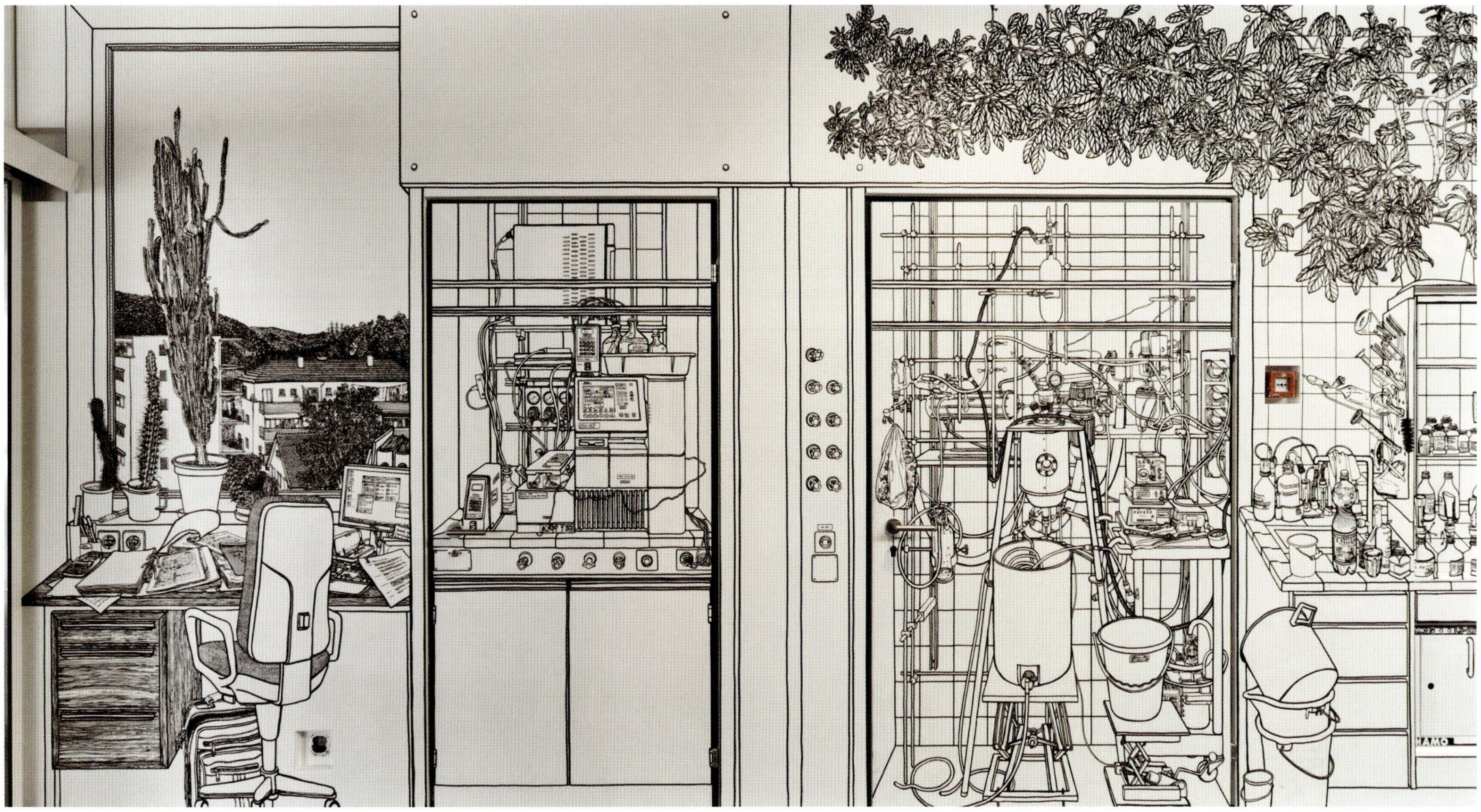

HAMO

GALLERY WALL DRAWINGS

This body of work is an ongoing part of Charlotte Mann's practice, currently encompassing wall drawings and printmaking. It investigates the relationships between domestic space and that of the palace, museum or gallery. She considers how we look at art compared to how we look at anything else and what the subconscious rules are that govern each person's gaze.

"I'm interested in making drawings of sections of walls in museums with extremely famous paintings hanging on them like the National Gallery in London or the Prado or the Louvre. Then to draw the wall exactly how it is: wallpaper, frame, information labels, electric sockets, humidity meters, paintings and all. The size of the work I make is dictated by the dimensions of a wall in a domestic space. This becomes a sort of cookie-cutter I use to take a chunk out of the museum wall, so the top of the painting might be cut off if it was something huge. The work becomes, amongst other things about scale and detail, and the nature of the place where my work is installed as much as the environment of the original painting and the museum it's seen in."

These two examples depict walls of the National Gallery, London. One is a screenprint which will be available to buy from Charlotte Mann's website, the other is a wall drawing on canvas installed in an apartment in Hong Kong.

Artist
Charlotte Mann

Location
Hong Kong, China

Client
Private client

Photographer
Titi Hui of Brick Lane Group of Hong Kong

THE SCHOOL OF LIFE CLASSROOM

The School of Life was founded in 2008 by philosopher Alain de Botton and Sophie Howarth, a former curator from Tate Modern, London. Sophie Howarth commissioned Charlotte Mann to create murals for the almost windowless basement classroom.

The commission was to engage with those involved in founding the enterprise and to create a thought-provoking and inspiring environment that would represent the school without doing this in a direct or dogmatic way.

Charlotte Mann says of this piece "I wanted to make something that would trigger surprising thought processes, like ripples of recognition in the viewer. Everything represented might resonate in its specificity with someone. I'm interested in the sensation of surprise and unexpected, unsought familiarity. I tried to imagine as broad a range of possible someones as targets for this. So for example objects depicted range from a specific pair of Nike Air Jordans, to bottles that an aficionado would recognise from the collection Giorgio Morandi had for making his still life paintings. The books on the shelves were chosen by all the different writers, philosophers and artists involved in founding the school. The whole process of making the work was a lot of fun as there were so many interesting people involved".

Artist
Charlotte Mann

Client
The School of Life

Location
London, UK

Photographer
Peter Mann

PETER JENSEN TINA BARNEY THEATRE OF MANNERS WALL

In 2006 Charlotte Mann created the backdrop for Peter Jensen's Spring/Summer 2007 show. The subject of the drawing was an invented interior wall comprised entirely of elements found in photographs form Tina Barney's book *Theatre of Manners*. Tina Barney was the Muse that inspired the Peter Jensen collection. Charlotte Mann, a fan of Tina Barney's work proposed the idea of the backdrop drawing to her friend Peter Jensen.

The drawing was thirty meters long, comprising three ten metre sections. It got a lot of press coverage and propelled this current phase of Mann's practice. She then created a related piece for the b-store gallery space on Savile Row in London. She said of the b-store piece: "This installation is something that has grown out of the initial backdrop drawing, especially that feeling of knowing the rooms of Barney's photographs so well. This knowing is made particularly tangible by the physicality of drawing life-size. Each object is handled in the imagination then set on paper according to the scale of the hand doing the drawing, and the space is measured by the body of the person who is doing the drawing: me. This time the subject matter had a more direct connection with particular photographs, the main inspiration being Ada's Interior. The figures that inhabit this space are taken from other photographs (The Son, The Cousins, and The Pink Lemonade) but instead of wearing their own clothes, they are all wearing items from The Peter Jensen Spring/Summer 2007 collection."

Charlotte Mann has since made two further installations for Peter Jensen shows: one based on Dianne Arbus' photographs, the other a faithful recreation of a school hall from the London primary school that Mann attended as a child.

Artist
Charlotte Mann

Client
Peter Jensen

Location
London, UK

Photographer
Chris Moore (catwalk shots),
Peter Mann

Architect

Cristina Parreño Architecture

Design Team

Cristina Parreño Alonso, James Coleman, Sharon Xu, Koharu Usui, Natthida Wiwatwicha, Hannah Ahlblad

Photographer

Luis Asin

PAPER CHANDELIERS

Paper Chandeliers is an experimental space where the simplicity of the materials appears as a counterpoint to the most advanced technology of fabrication.

Paper Chandeliers is a big undulating roof made out of paper tubes that illuminates the spatial environment mediating between the art installation and the architectural project.

The project was selected to be built for the yearly International Art Fair ARCO Madrid 2013 and was supported by MIT Architecture.

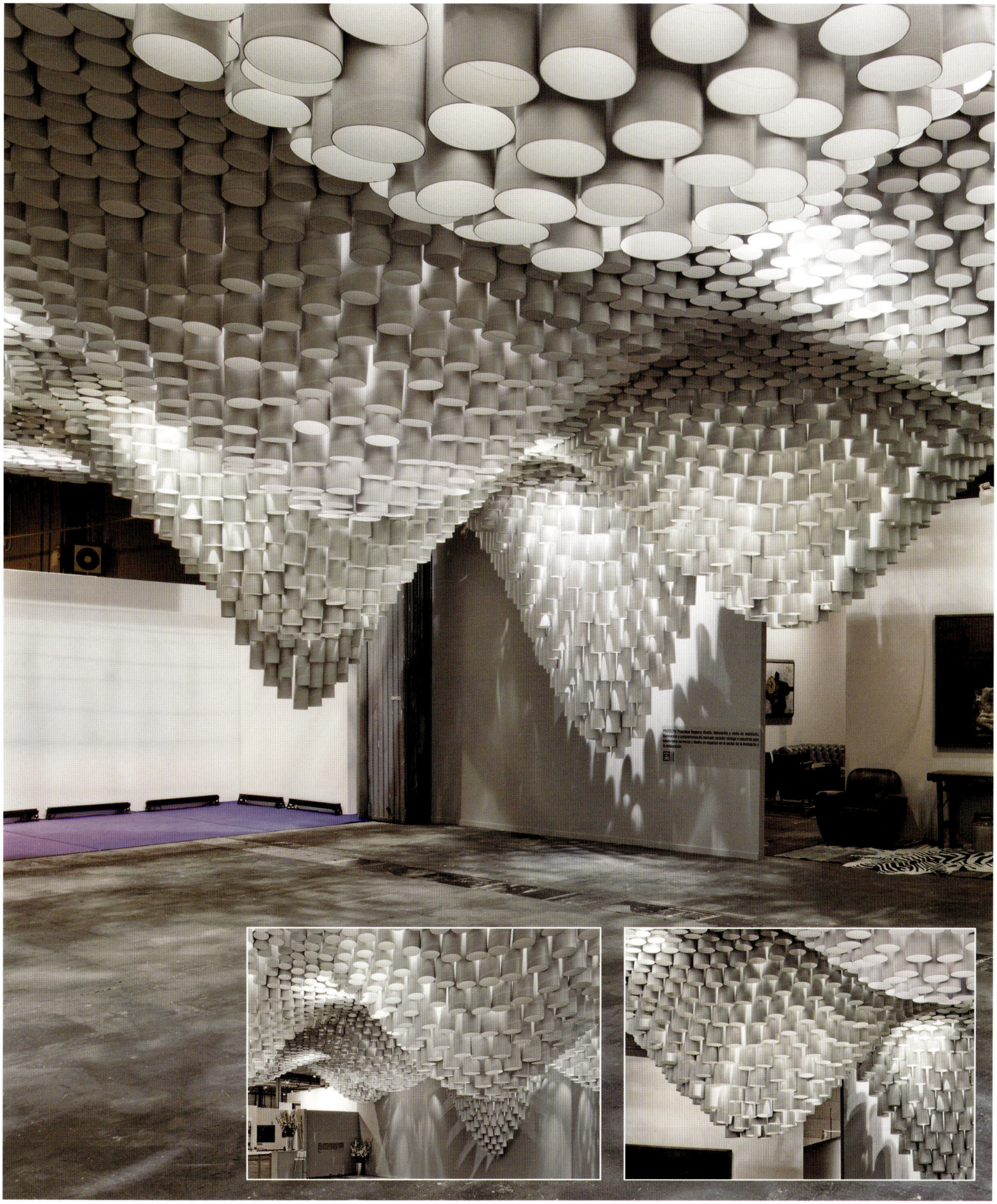

PICCINO

+Quespacio just finished a new store in the City of Science, Valencia, called Piccino. Piccino is a store that brings kids clothes from Italian brands to the city of Valencia, like Brems and Bimbus.

Giving major importance to the clothes exposed in the store, the interior space is left mostly white. However 2 kids, inspired by the store-owners kids, welcome customers and their children with some playful and fresh colors. Kids, part of this sewing clothes shop, in which they work hard designing clothes for their new friends.

In a parodic way old furniture is represented by vinyls that combine shelves and actual furniture with silhouettes that are making fun of classic furniture. In the meantime the fresh colors used to represent the kids' work are getting back the modern touch to the store. Some magnetic pixels are giving the chance to the children to create their own clothes while others are watching an animation movie Last but not least; the classic Lou Lou Ghost chair from Kartell seems to be made for this store.

Piccino invites you all to their new, fresh concept store bringing you their sewing clothes shop loved by young and old with an eye for little details.

+Quespacio in this project worked out again their philosophy of creating corporative projects using creativity, taking charge of the branding and the interior design of Piccino.

Piccino

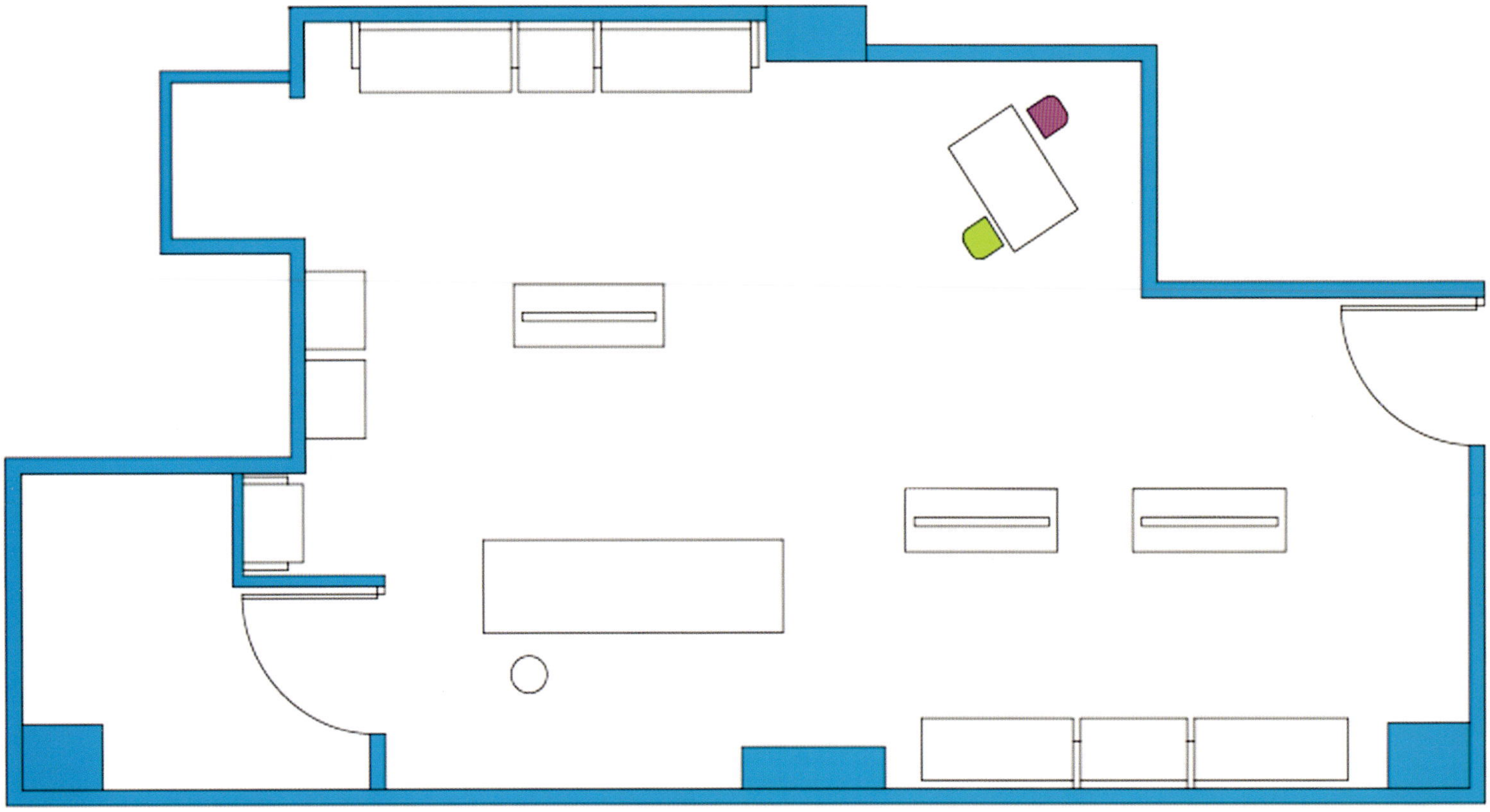

Calle de la Rosa
Calle de Trafalgar
Calle Luis Bolinches
Calle Pintor Maella
El Corte Ingles
Calle de Menorca
Calle de Menorca
AQUA
Encuentranos en
Trafalgar 52, Local 5,
46023, Valencia
contactenos@piccino.es
Tel: 963327840
www.piccino.es

Piccino
Ropa italiana
para niños.
Brums
BIMBUS
Ven a
conocernos,
te gustará!!!

Design Agency
Masquespacio

Designer
Ana Milena Hernández Palacios

Graphic & Interior Design
Ana Milena Hernández Palacios

Client
Piccino

Location
Valencia, Spain

Area
38 m²

Photographer
David Rodríguez

BISTRO JAPONAIS KINOYA

The designer has given a humoristic twist to the traditional "tatami" room which is at the heart of the project. It breaks the "cafeteria style" effect which is due to the high ceilings and the rectangular shape of the place and forces us to question our perception of space. The long sculptural strip made of yellow birch wraps itself around, as would a furoshiki, to form a "box" that the designer has positioned slightly off the mark. The reason for this becomes apparent when one moves along inside the bistro bar. In this clean space, however, not a single detail is superfluous.

The gigantic flowers that run on the floor, walls and ceiling lead us into a world of fantasy and poetry. Such excessive demonstration is often seen in a manga universe. Here, the huge plum blossoms, peonies and stylized chrysanthemums have become contemporary representatives of the kamon (the heraldic insignia of ancient samurai clans). These identical patterns are also found on a tiny paper kimono displayed on the wall, the first of a series of ten works of art, which shapes and colors are a daring challenge to one another. This quintessential signature has become a part of the identity of Kinoya, so much that the aesthetic floral graphic is also displayed on the menu card.

The warmth and golden quality of wood, the delicate and sharp shades of pink against a masculine glossy black; the softness of materials in a more classic shade of grey all blend in with finesse, because for Jean de Lessard: "I wanted to create a place which would be sexy but not fake, a place where both men and women would feel good." Still, humour is never far away evokes the playfully clad-in-skirt bar stools created by the designer. Soft and indirect lighting is used to create different moods, as well as the play of chiaroscuro and of pitching "full" against "empty" through various architectural elements, such as the wood partition at the bar that the designer has detached from the wall.

Designer
Jean de Lessard

Client
Kinoya

Material
paint, yellow birch, laminate, laquer

Photographer
M David Giral

LJUBLJANA PUPPET THEATRE

LGL is the oldest puppet theatre in Ljubljana, located in a classicist building from the turn of the (previous) Century. It caters mainly to children, but also produces many performances for adults. Still the lobby needed an update. The designers created a fantasy world, a sort of a mixture of Willy Wonka inspired psychedelics and puppet-like characters with a bit of the Wizard of Oz thrown in the mix.

Designer
Kitsch Nitsch

Client
Lutkovno gledališče Ljubljana (Ljubljana Puppet Theatre)

Location
Krekov trg 2, Ljubljana, Slovenia

Use
Culture / Theatre

Area
250 m²

Photographer
Kitsch Nitsch

Main Material
wall vinyl stickers, faux marble linoleum flooring

VELIKI ODER
MALI ODER

YMS CENTER HAIR SALON 1ST & 2ND

They were commissioned to find a visual style that goes as far away from the existing brands as possible and teens and people in their early twenties can relate to. They wanted to avoid the main mistake that is usually made when designing for youth, where one takes the symbolism of a subculture and tries to buy acceptance by photoshoping it all over their own image. Such attitude always comes out pushy and patronizing and you rightfully get just the opposite result. So they turned to their favorite source of inspiration, the 70s and 80s brilliance of postmodernism that has a wonderful way of speaking against functionalism with the use of strong graphics and over the top furniture. It's nonconformist, just as youth is (or at least should be) and that was the common ground they could relate to. They also felt that by introducing a style that is directly nonexistent in any of the teen subcultures right now, they can equally speak to all of them. More importantly they tried to conceptualize a perfect hairstyle salon, which is in its essence an overproduced and over designed place, a bit fake and with appearance overweighing the substance — but with a lot of human warmth and affection. Well, it was either that or decorating the walls with nailed up skateboards that someone suggested they should do. They believe they made the right choice, since none of us skate.

Since the beginning of the project 2 salons were opened, the first one in the centre of Ljubljana and the second one in a vast shopping centre called City Park on the outskirts of the city.

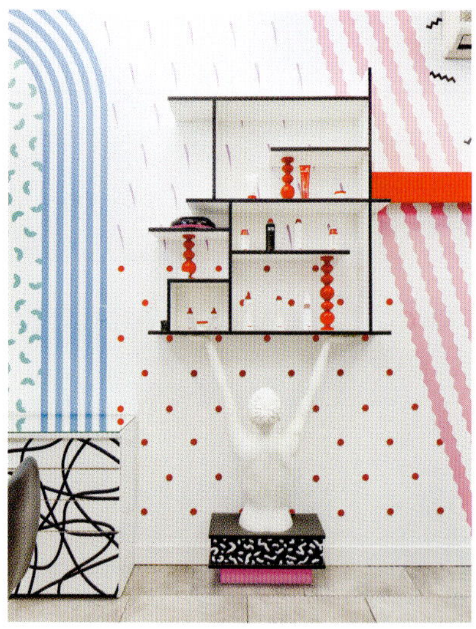

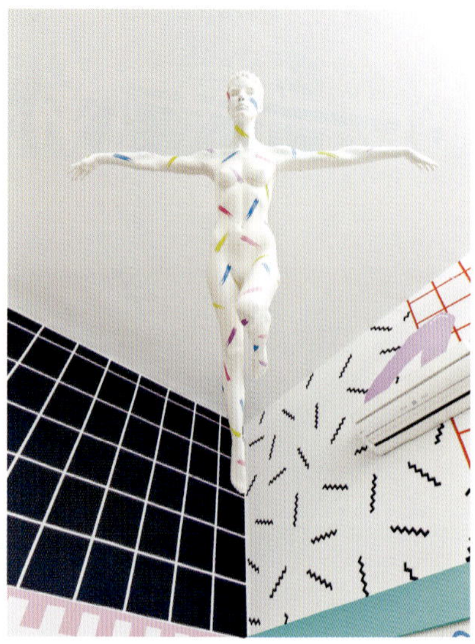

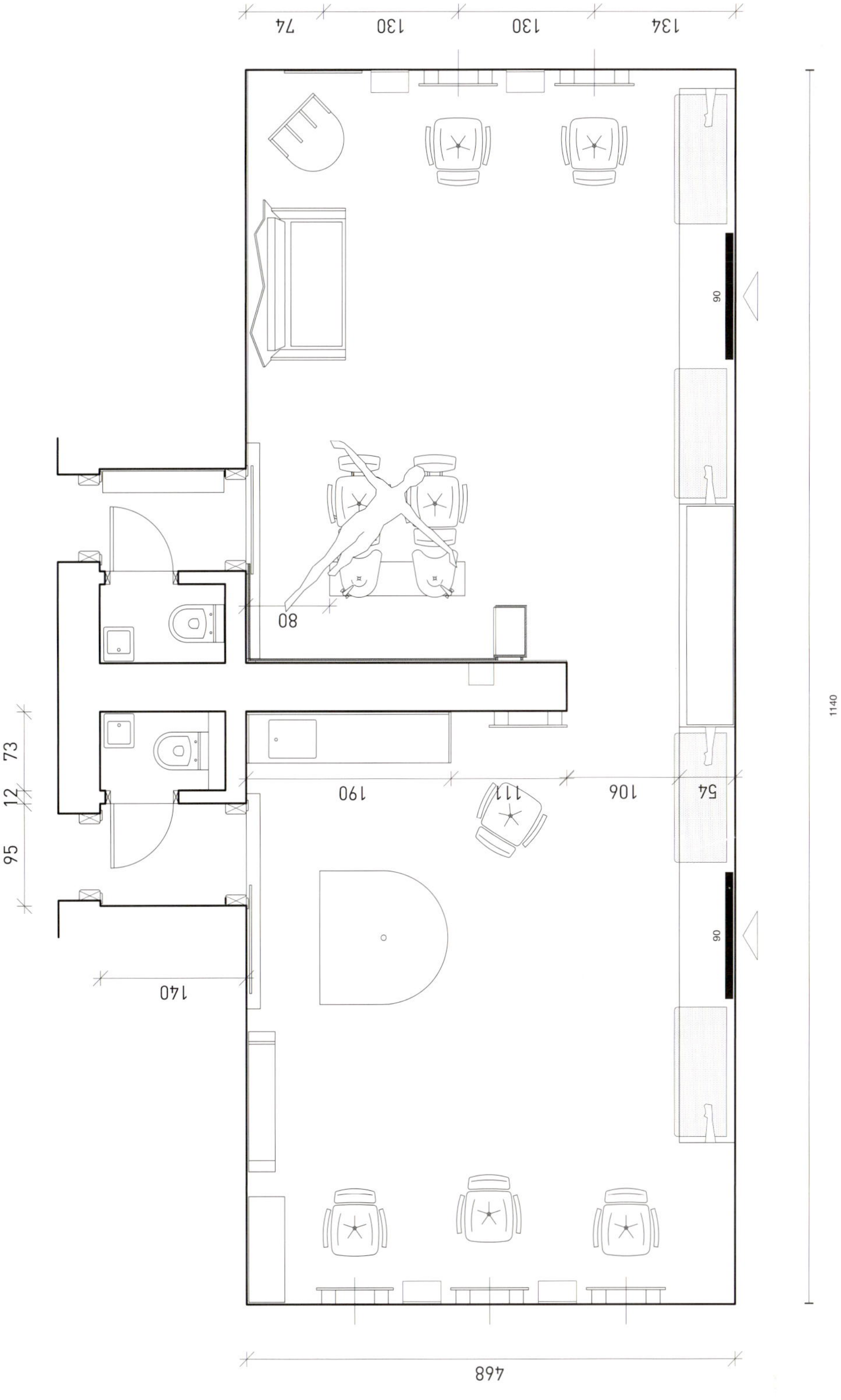

Designer
Kitsch Nitsch

Client
Mič Styling

Location
Ljubljana, Slovenia

Photographer
Miha Brodarič / Multipraktik, Kitsch Nitsch

Main Materials
- wall vinyl self-adhesive film
- painted MDF wood panels for construction of furniture and mirrors (all furniture was designed by Kitsch Nitsch and made by Lavka Carpentry)
- ready made mannequins

Car Wash
Military Cut
Red Carpet
College Classic
Southern Comfort
Barbie Blond

"ATTENTION! PRIVATE SPACE." – A TEENAGE ROOM

The years when a person wants to have for the first time his own private space are in the teen ages. During this time the young person has a lot of physical and mental changes.

The room we created is for a 14-year-old girl who is interested in art, especially in the graphic design.

She believes that she is very different from the other teenagers and that she will be a famous graphic designer one day. Her parents want to encourage her. They gave designers the privilege to design her space.

The designers divided the room into several zones according to their functions. Every zone has different feelings. Through the use of colors they transformed the usual forms of the room and the furniture into different shapes which have an impact on the thought process of the girl who loves art.

The main function of the space is to develop the girl's imagination and to give her more freedom to think and to dream.

Design Agency
Gemelli Design

Designer
Branimira Ivanova & Desislava Ivanova

Design Agency
Gemelli Design

Location
Bulgaria

Designer
Branimira Ivanova & Desislava Ivanova

Gross Area
40 m^2

AN OASIS IN A SANDSTORM

Gemelli Design studio designed a bedroom and a bathroom separated from each other with a fireplace and a glass wall.

The main objective of the spaces is to provide a complete relaxation of senses — in the bedroom the color of the imagination and spirituality — the purple; Yellow is used for the bathroom. In this way the bathroom is like an oasis in the desert.

They chose two different types of sinks, one small round and the ocher wide long, which can be used for different purposes. They decided to use console bidet and standing toilet. The bath is located next to the window with a nice view to the sea.

The bathroom favours quiet conversations. While one is taking a bath, another one is having a cup of wine on the sofa enjoying the fireplace.

H_2O IN GEOMETRY

The bathroom is a very important room in a house and is probably the only one place for staying alone, relaxing, thinking and dreaming.

The designers chose the name H_2O in chemistry, because water is a leading element in the idea of their bathroom project. It represents the natural connection between this room and the man.

According to the Japanese researcher Masaru Emoto water can memorize all effects of the surroundings, including our thoughts, changing its structure. Then, with these new properties, water can affect them.

All known forms of life depend on water.

The bathroom is divided into two areas at different levels with specific functionality. Despite of this separation the designers aimed to design the bathroom as an entire organism. The wash-basins, the toilet and the bidet are located at the lower level while the shower, at the upper level.

The two levels are connected by stairs (which can also be used for sitting).

Each area is determined so as to give freedom of the body and the spirit.

In the zone around the wash-basins, behind the mirrors there are niches in the wall which can serve without obtrusion. There is a storage cupboard between the wash-basins. The sanitation area also has a niche which provides a place for books and a small rolling lap top.

On the second level are separate shower and tub, where you can relax.

Design Agency
Gemelli Design Studio

Designer
Branimira Ivanova & Desislava Ivanova

Gross Area
12 m^2

CHILDREN SHOES EXPOSITION

Masquespacio designed the exposition of children's shoes for the Spanish International Fair of Children's Fashion (FIMI) celebrated in Valencia from the 25th until the 27th of January 2013. For this 76th edition FIMI organized an exposition to make a tribute to Spanish children's shoes brands, counting with the participation of 39 national brands among which were included some brands of international relevance like Gioseppo Kids, Conguitos Goflex and Garvalin.

The exhibition designed by Masquespacio is traduced in a forest of autumn flowers realized of foamboard and with fashion patterns. Ana Milena Hernández Palacios comments, "As FIMI mostly is a fashion fair and not a shoes fair, to every flower is applied a bunch of patterns inspired by fashion. That way we wanted to better integrate the shoes brands between a huge amount of clothes exposed." The 39 flowers that were hand cut in foamboard contain different patterns and forms that make them flourish like unique, presenting a pair of shoes from every participating brand. Entering to the exposition, corridors were created that guide the visitor across the exhibition, while he enjoys the best Spanish children shoes brands. Thanks to the suspension of the elements with a nylon thread, the flowers twist slowly showing every side from the pairs of shoes exposed. In the trees provided by FIMI in the meantime they can know the legends behind every brand. In this manner the visitor experiments a moment in which he can disconnect from the rest of the fair, crossing a forest of flowers inspired by autumn and fashion patterns, enjoying a selection of the best Spanish children shoes brands.

Masquespacio with this project desired to demonstrate again creativity has no limits and that it can be a strong factor to sustain visually any product, making it more attractive for its potential buyers.

Briefing about materials and exposure requirements: "We needed to work with materials available or producible by the fair itself to make the exposition profitable and reduce costs. In first instant we visited the warehouse of the fair to take a look at elements used during previous fairs. As we didn't find immediately enough elements to create a visually attractive and coherent unity, we decide to work with foamboard and vinyl, two materials that could be produced internally. The different elements (flowers, clouds and circles) were designed in illustrator by us and hand cut by the fairs signmakers. The trees, in which we find the legends of the brands, where created by FIMI during their previous editions. It was a requirement from the organization to include them in the installation."

Design Agency
Masquespacio

Designer
Ana Milena Hernández Palacios (Graphic design and exposition)

Photographer
David Rodriguez from Cualiti

Client
FIMI (Feria Internacional de Moda Infantil)

Area
30 m²

Materials
Foamboard, nylon thread and vinyl

CALZADO
INFANTIL
Español

ARMSTRONG FAIR STAND BAU 2013

Design Agency
Ippolito Fleitz Group – Identity Architects

Design Team
Peter Ippolito, Gunter Fleitz, Tim Lessmann, Tanja Ziegler, Alexander Assmann, Sungha Kim

Stand Construction
ARTec Messebau

Client
Armstrong DLW GmbH

Location
Munich, Germany

Area
155 m^2

Photographer
Armstrong/P.G.Loske, Ippolito Fleitz Group GmbH

Munich's "Bau" trade fair is one of the world's leading trade fairs for the construction industry. Its particular focus is on building materials. For the "Bau 2013", Armstrong DLW is highlighting its product palette of linoleum and vinyl flooring. The idea is to pursue the twin goals of presenting its new collections and sensitising a younger and inherently more experimental target group of architects to the appeal of these materials. The stand exhibits both the Linea collection, featuring a wide spectrum of wood grain effects, and the Colorette product line, which is distinguished by strong, rich colours.

The Armstrong exhibition stand functions as a communication platform and gives visual expression to the company's field of expertise. Every surface is covered by a complex, geometric pattern consisting of different cuts of Armstrong materials. A large rear wall, concealing several support rooms, carries a striking, abstract interplay of colours and shapes that create a sense of depth and perspective. From this starting point, the spatial graphic spreads out across the entire floor, covering the reception counter and conference tables. Only an elongated counter directly in front of the rear wall, which is demarcated as the principal zone for discussion by a firmament of pendant lights, remains entirely white. An open communication zone of tables and counters is demarcated and contained by a folded ceiling element. The latter serves as a three-dimensional counterpart to the two-dimensional structure of the spatial graphic. A polygonal podium encloses a calmer communication zone within the stand. Here different materials are presented on the steps and retractable side wall elements. An unconventional note is added by the fact that visitors are invited to take a seat on the steps themselves, creating a concentrated yet relaxed room for discussion.

The Armstrong exhibition stand makes a strong and striking visual impact that can be transported well. Using an installation that encompasses the space in a collage of materials, Armstrong recommends the construction material linoleum for use in contemporary, cutting-edge interior design.

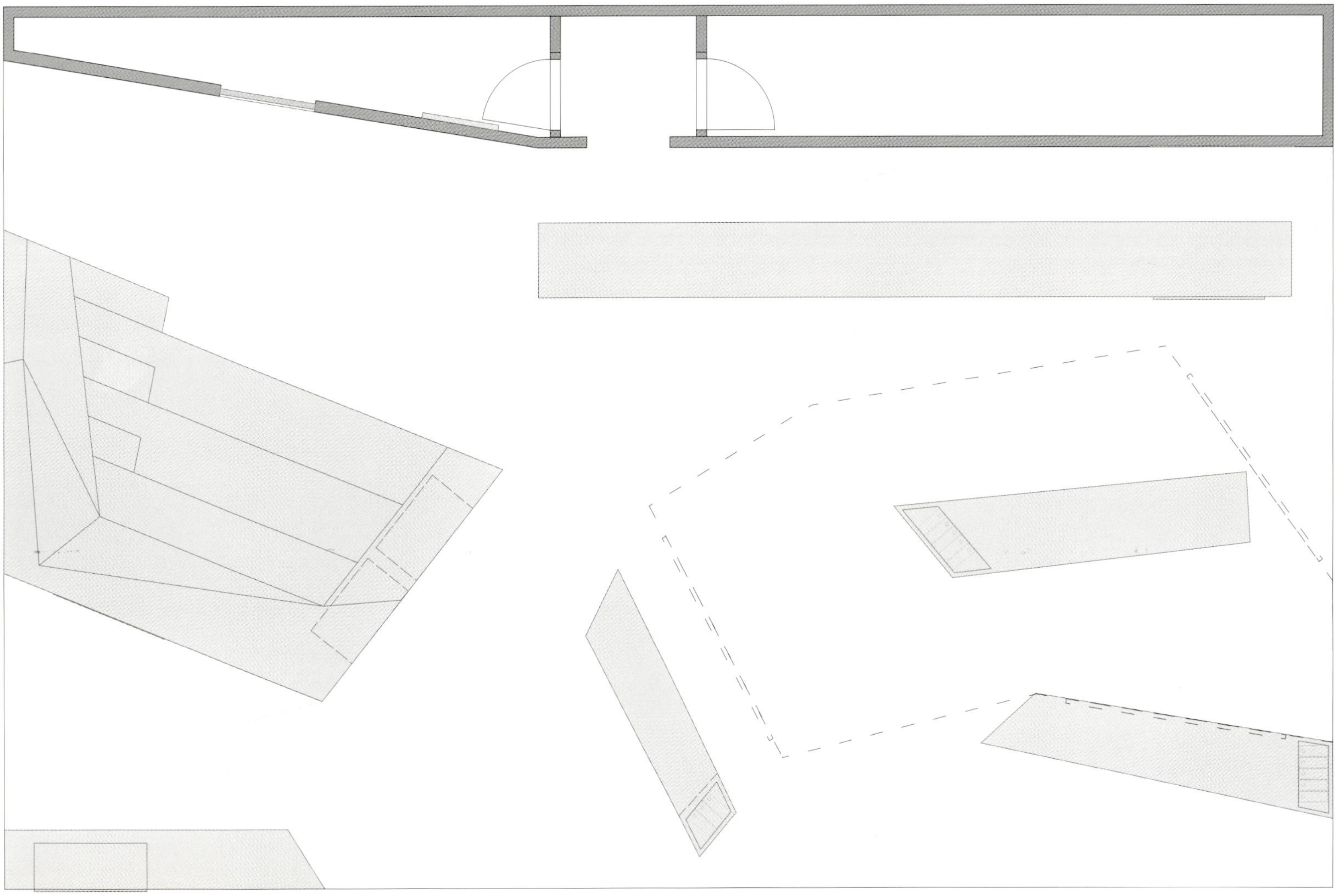

ürliche Rohstoffe für ein ökologisches Produkt
Natural raw materials for an eco-friendly product
LINOLEUM

Nachhaltiges Bauen mit natürlichem Mate
Green buildings with natural mate
LINOLEU

EUM

Armstrong
FLOORING

TAIWAN NOODLE HOUSE

Studio Golucci International Design has completed interior design for the noodle restaurant, located in China. The walls are completely decorated with authentic plates that creates unique atmosphere.

The Taiwan Noodle house located in Ningbo China is not only a place for delicious authentic tasting noodles but also provides its customers with an unbelievable dining experience that brings them back to their childhood. The stylish and clean design done by Golucci International Design is intended to represent all of the elements of an old school noodle house, the open space, large porcelain bowls, plenty of chopsticks and the hot steam that always filled up the place. The single large dining room encourages customers to socialize and enjoy a quality meal with family and friends. The thousands of plates used to decorate the wall create a reminiscent mood; helping customers understand the sole purpose of this restaurant, which is all about the celebration of memories and tasty noodles.

The Taiwan Noodle House restaurant interior by Golucci International Design in Beijing, China, is a nice mix between contemporary design with Wegner chairs and Tom Dixon pendants on one side and traditional noodle bowls covering a full wall on the other.

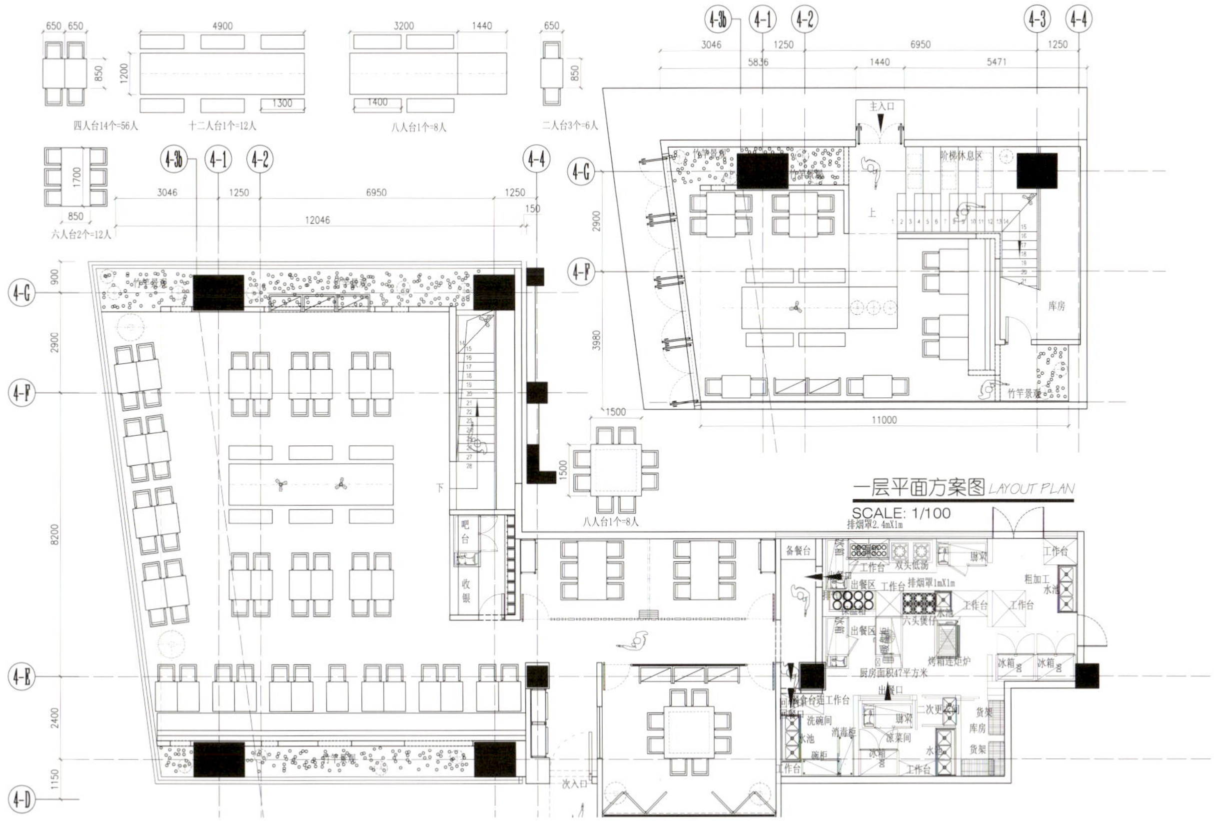

Designer Agency
Golucci International Design

Design Team
Lee Hsuheng, Zhao Shuang

Client
Taiwan Noodle House

Location
Ningbo, China

Built Area
350 m^2

Photographer
Sun Xiangyu

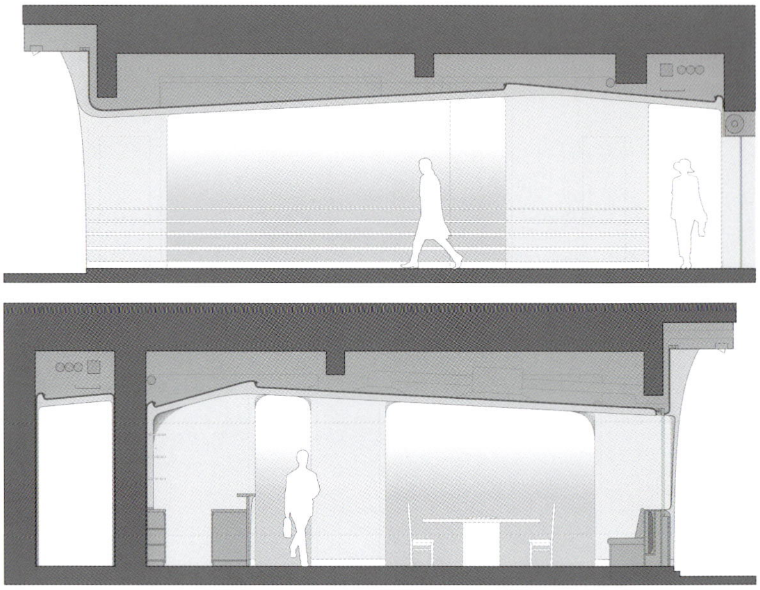

PAK LOH CHIU CHOW RESTAURANT

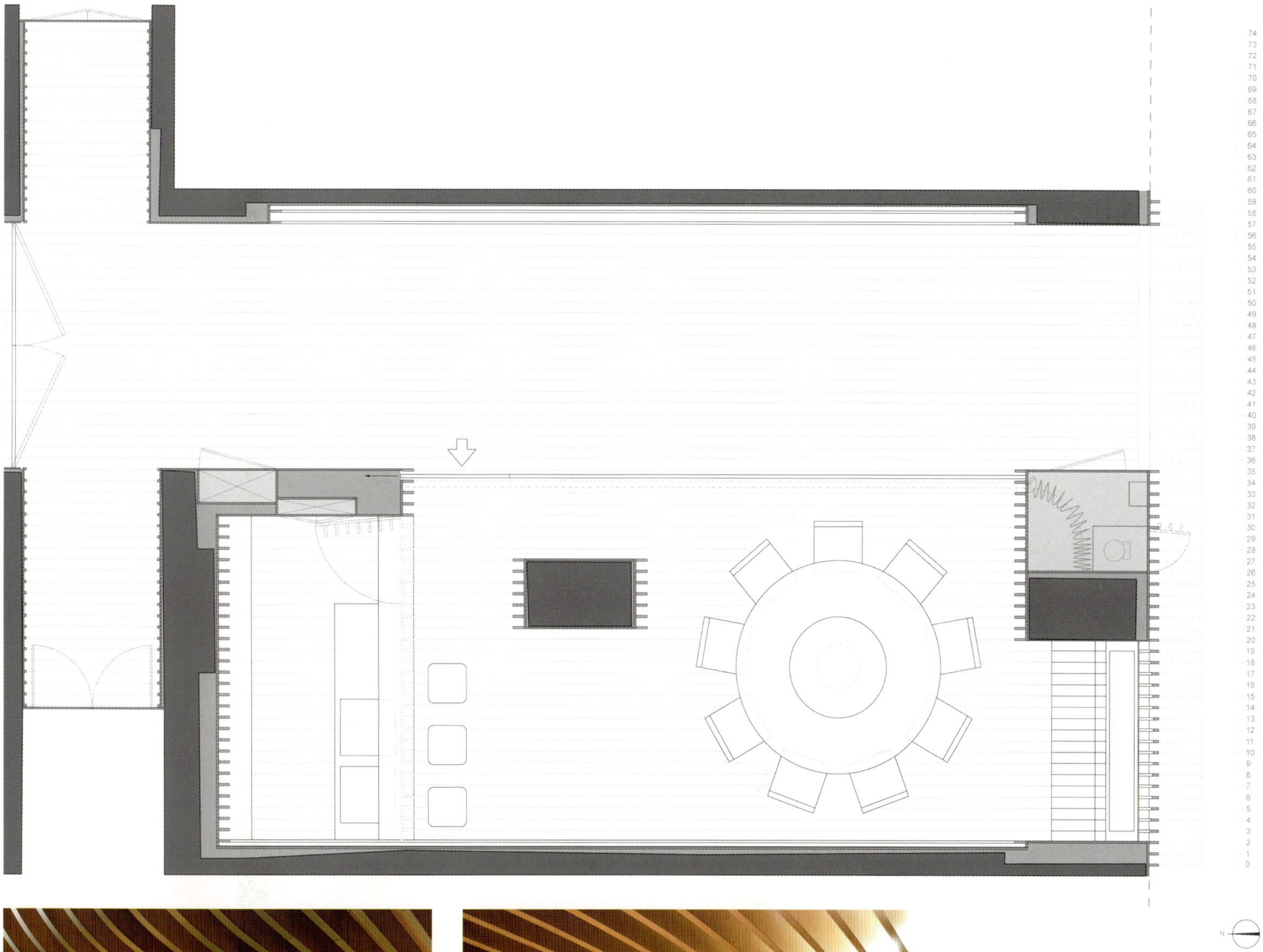

Design Agency
Laboratory for Explorative Architecture & Design Ltd. (LEAD)

Design Team
KristofCrolla, Sebastien Delagrange

Project Management
4N Architects, Laboratory for Explorative Architecture & Design Ltd. (LEAD)

Client
Pak Loh Chiu Chow Restaurant Ltd.

Location
Hong Kong, China

Total Area
52 m^2

Photographer
Dennis Lo Designs

Pak Loh is a traditional, Hong Kong based family restaurant specialised in Chiu Chow style Chinese food. For over 45 years its highly successful flagship restaurant has been located in the heart of a building block in Hong Kong's Causeway Bay district. This project involves the renovation of the restaurant's public entrance area and the replacement of its front Lo Mei food preparation area with a high-end VIP dining room / lounge area.

The new VIP room is located in front of the restaurant at the building block's façade, and is disconnected from the restaurant's main entrance by a public passageway. Thus, this very small and intimate private room is fully surrounded by publically accessible space. This condition was grasped as an opportunity to engage those passing by in a unique and playful visual theatre: all designed elements work together to create an interaction between Causeway Bay's busy street-life and the restaurant hidden deep inside in the building block. This happens by working with a continuity of the space's geometrical elements, by amplifying perspectival views and by a play with transparencies and reflections. Through the subtle revealing of hints of the VIP dining experience to the outside, potential clients are seduced and enticed to enter the space.

The main feature of the project is its series of 75 parallel, continuously undulating, curvy bamboo fins. These are combined with parallel mirrors on opposing walls, which visually enlarge the very narrow passage space and create a dynamic perspective that draws visitors' gaze towards the restaurant's main entrance. The geometric challenge of the project reveals itself in the subtle sectional variation of the dynamic space. Here one can see the gradually changing continuous curvature of the fins, of which the fronts create a flat, single and double curved surface. All fins are CNC (computer numerically controlled) milled from a flat sheet and can thus easily have accurately cut curved edges. The back surfaces on which these fins were mounted are also made from flat sheet material, which has been grooved in particular areas to allow it to be bent. This method reduced the geometric possibilities there to single curved cylindrical or conical surfaces. These complex surfaces were strategically used only in the corners of the room, where they were combined with built-in lighting. Inside the VIP-room the back-surfaces are finished with a dark blue fabric, which contrasted starkly with the light colour of the bamboo finishing of the fins, allowing the geometric patterns to emerge more strongly. The separation of discreet building elements as columns, walls and ceiling is abandoned as the undulating ceiling blends in fluidly with columns and walls through slumped edges of which the effect is amplified with the use of mirrors. Structure, lighting, television, and air-conditioning are built into the fin-system to provide an undisturbed spatial continuity. The concealed window behind the façade allows a filtered connection of the VIP room's interior with the street, while the corridor window with gradient transparency print provides privacy to the diner party while the ceiling pattern is allowed to continue. Concealed, heavy, dark blue curtains allow the room to be fully enclosed from curious views.

The elegance of the curves, the disappearance of space-limiting elements such as corners or walls through the use of mirrors and fins, the seductive game of exposure and concealment, all work together in giving the VIP room a status of exclusivity. By revealing glimpses of this experience, the space is to trigger curiosity with those who catch a glimpse of what happens inside this intimate, luxurious private room.

FARMA KREATON RESTAURANT

Restaurant "farma kreaton" is located at the back yard of a two-storey building at the centre of Komotini in Greece, and is the direct extension of the "FABRICA KREATON" restaurant.

The main goal of this project was the creation of a scenic that brings to mind an actual farm. At this particular farm the imaginary, which is represented by the fairy-like digital printings, and the real, which is represented by the users-clients, co-exist. The "farma" abounds with elements such as forms of the domestic animals, materials in their initial form, proximity and comfort, with apparent features from the countryside ecologically approached.

The facade of the restaurant-bench of the farm is decorated with piles of straws and colorful blossoms placed into tin boxes. Entering the restaurant two friendly cows welcome people in their farm and lead us to the main roofless corridor. Here the barrels from the cellar play the role of the chairs and tables, while metal tanks hanging upside down illuminate the corridor. On the right and left sides of the restaurant have been constructed two separate roofed areas with big tables and long benches with a view at the meadow and the stores, where proud horses stick up from their barns and kind sheep tour us around the "farma". What's more at the end of the main corridor is the imposing coop, which houses a big dining table apart from the hens and their nests.

Design
Minas Kosmidis (Architecture In Concept)

Location
Komotini, Greece

Area
287 m^2

Graphics
Yannis Tokalatsidis

Lighting Design
Minas Kosmidis

Photographer
Studio VD, N. Vavdinoudis – Ch. Dimitriou

HATCHED AT HOLLAND AVENUE

Design Agency
Outofstock Design

Location
Singapore

Photographer
Kim Jung Eun

Following the popularity of the first Hatched restaurant, Singapore and Barcelona based design collective Outofstock was recently commissioned to design a second restaurant at 267 Holland Avenue, Singapore.

Located in a two-storey historical shop house in Holland Village, the design of this egg-themed, all-day breakfast restaurant is based on the concept of a nest. The intention behind the design was to inject more colour and playful elements into the new restaurant, building upon the warm and cozy barn house atmosphere of the original establishment, also designed by Outofstock.

The nest facade was realized with yellow braided rope woven around a steel frame with laser-cut holes. This steel frame also holds up the glass panels and a floor-sprung rough-sawn timber door.

The restaurant uses mostly original lighting and furniture designed by Outofstock, such as the Naked chair produced by Bolia, the Biscuit stool and table produced by Environment.

The bar counter is composed of offcuts from teak wood floorboards. The floorboards were used to compose a herring bone pattern, hence the almost perfect triangular offcuts.

Abstract wall lighting fixtures which could be interpreted as hatching eggs or peeled potatoes act as conversation pieces on the upper floor of the restaurant.

HATCHED
@HV
TUES - SUN
0800 to 2200
Tel: 6463 0012

EMILY COFFEE SHOP

Located in the upcoming district of Sanlitun the new branch of Emily Coffee Company features a combination of warm tones, and organic materials which seek to set a balanced stage for a rich coffee experience.

The design in contred around three elements; the adaptive wood skin in the perimeter of the shop, the lighting design and the choice of materials and colours palette for flooring, walls and counter.

The intention behind the wood skin is to blur the existing edges and corners to create a continuous, visually uninterrupted space.

The driving shape behind the wood ribs stretches around the space and blends with the planes. The resulting wood grid provides shelving space for product advertisement and its depth enhances acoustics by absorbing excess of sound waves.

All wood ribs were set out digitally and cut using CNC technology. The horizontal and vertical ribs have male and female sluts. This type of assembly minimizes the use of toxic adhesives and mechanical elements.

The lighting design includes a combination of heavy duty industrial lamps for the main counter, led directable stage lights targeting the menu and products and vintage carbon filament bulbs for the sitting areas.

As for the materials, the floor tiles are 800 x 800 metallic glazed porcelain tiles. The main counter is finished with white car paint and routed on fillets. The wood was harvested from recyclables sources in Heilongjiang province in the North East of China.

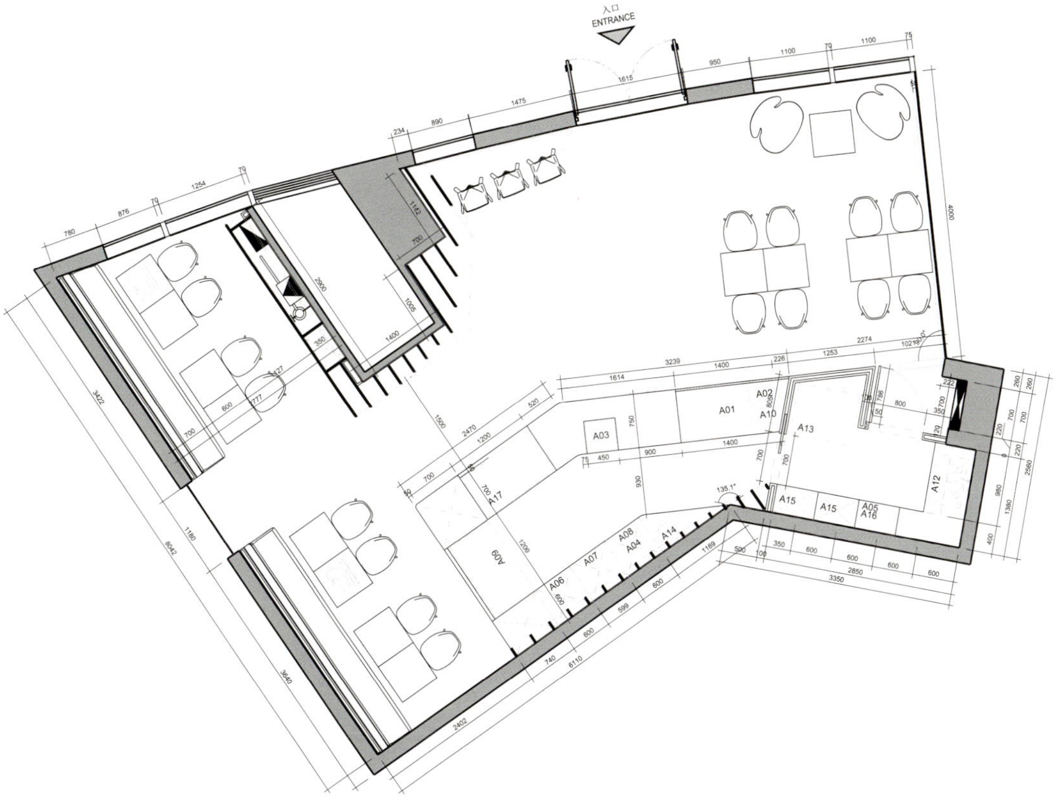

Design Agency
AIDIA Studio

Project Team
Rolando Rodriguez-Leal, Natalia Wrzask

Location
Beijing, China

Client
Emily Coffee Company

Area
68 m^2

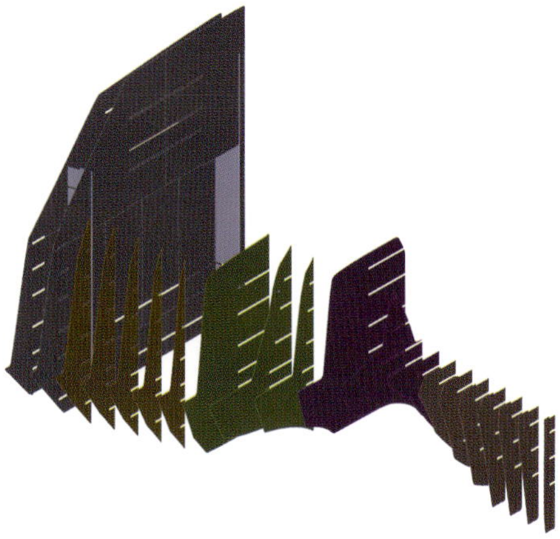

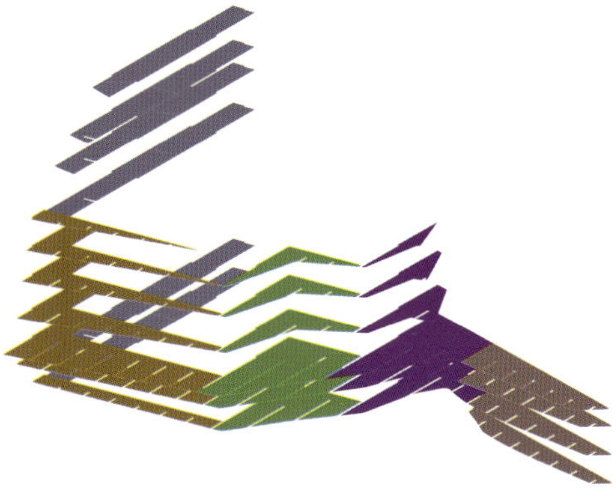

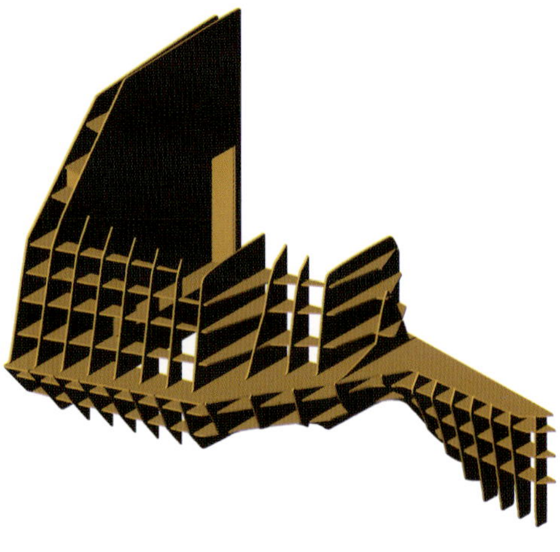

Emily

OLIVOCARNE RESTAURANT

It was designer's intention, when designing new Mauro Sanna's venue in London (now the fifth one the designer designed for him), to narrate about Sardinia (Mauro's as well as his home island) through iconographic references to the main points of its traditional economy sheep farming and handicrafts (weaving, in this case) and the quotation of the works of a Sardinian contemporary artist, Eugenio Tavolara, who remarkably contributed, along his whole lifetime, to bring out and safeguard our traditional culture. Far from wishing to evoke Sardinia through trite images good for low cost turism, the designer's has been told with a language which winks at contemporary design, also resorting to the work of some skilled Sardinian artisans, like sisters Stefania and Cristina Ariu who have moulded a huge terracotta made bas-relief evoking a flock of sheeps, and Mauro Angius who has given life to a crowd of peasants, horsemen, shepherds, wild boars and hunters which animate the restaurant's walls, in a sort of imaginary gallery evoking Sardinian country life.

Design Team
Pierluigi Piu
Ceramist sisters Stefania & Cristina Ariu (terracotta bas-relief), Craftman Mauro Angius (Corian silhouettes spreaded on the walls)

Lighting Designer
Pedro Pinto

Photographer
Riccardo Sanna, Pierluigi Piu

Location
London, UK

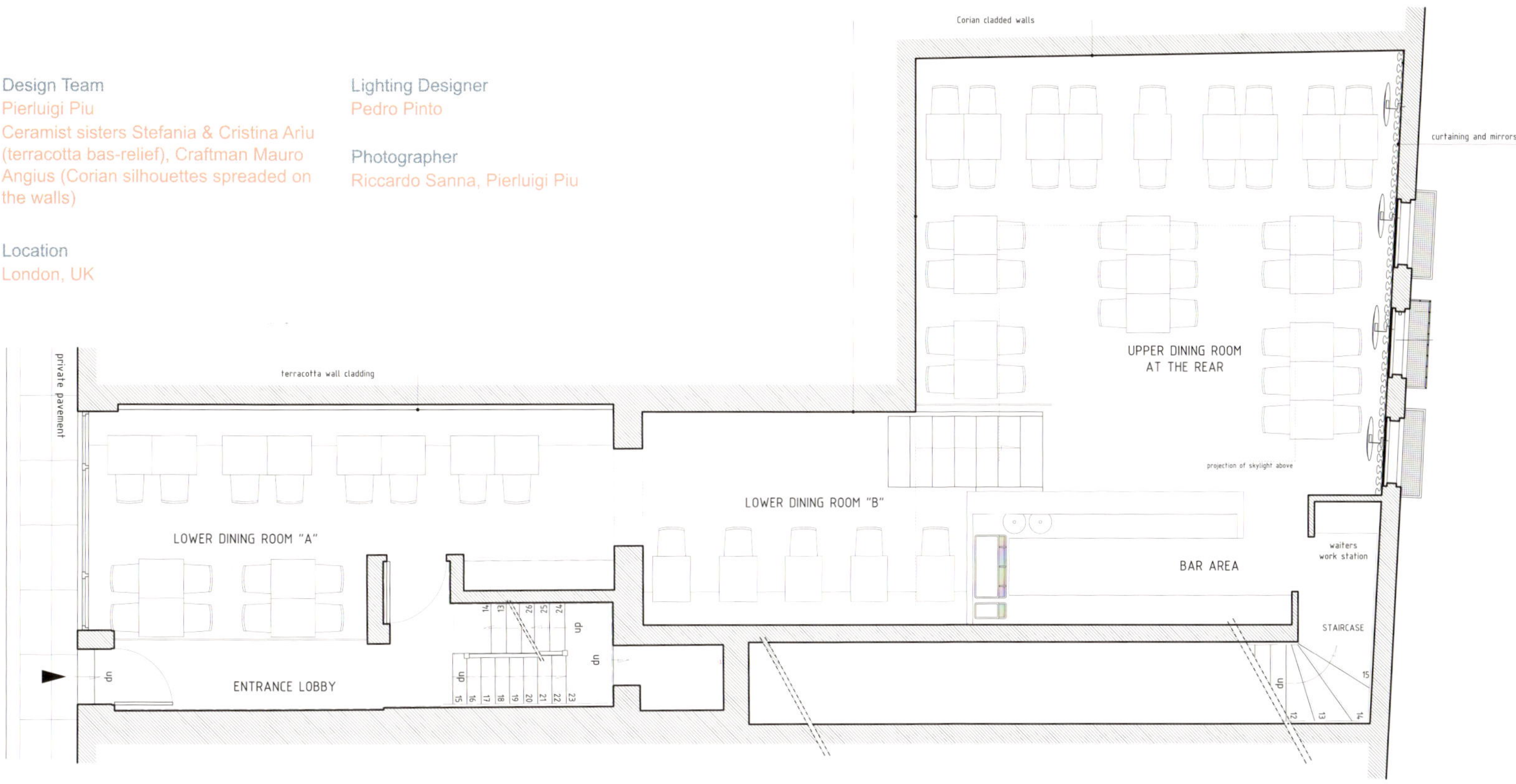

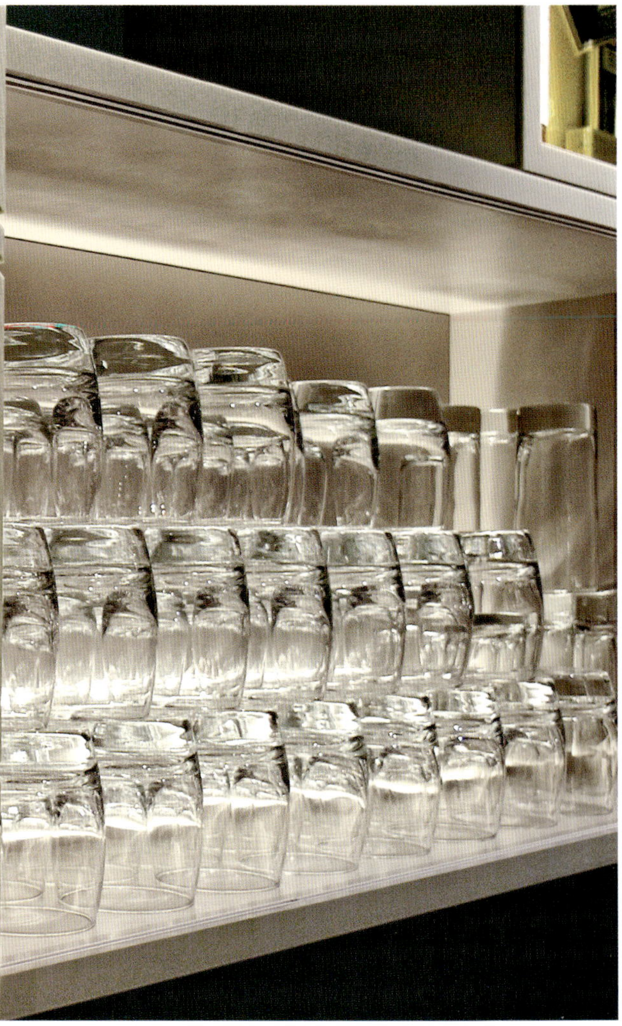

DISARONNO
BAILEYS
PIMM'S
MARTINI

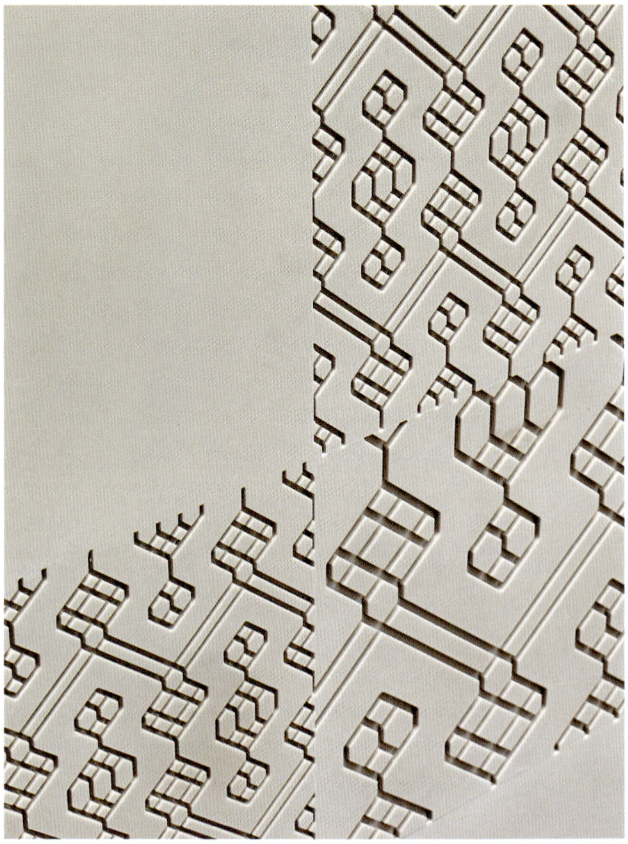

MINI POP-UP STORE

Not only Olympia was the center of attention in London in the past – also the new MINI Pop-up Store in one of Europe's largest shopping malls, London's Westfield Startford City, is still attracting much attention since its opening in spring 2012.

Most recently in July 2012 the MINI Pop-Up Store Project was awarded the renowned red dot design award 2012 in the category "Event Design". Together with the client MINI, the Berlin creative agency for retail communication, Studio 38, can also feel proud that their work came out on top in a field of 6,823 entries from 43 countries. The red dot jury, made up of international design experts, was convinced by Berlin-based company's pioneering and smart designs.

Studio 38 were responsible for the overall concept and implementation of the MINI Pop-up Store, from the architecture to the visual merchandising and lighting to the finer details such as the wall graphics and olfactory and sound design.

MINI
MINI
JOHN COOPER WORKS
DARE TO BE DIFFERENT.

MINI
MINI
JOHN COOPER WORKS

MINI
MINI

MINI
JOHN COOPER WORKS

Design Agency
Studio 38

Architecture
Mark Bendow – Interior Environments

Project Manager
Kathrin Janke-Bendow, Yannah Bandilla, Reinhard Knobelspies, Mark Bendow

Photographer
diephotodesigner.de

SWINGING LONDON
SWINGING LONDON
TOUCH
YE61 OTF

MINI
Beat THE Street
MINI

TOUCH

EXTREME
RACING
37
MINI

MINI
MINI
MINI

EXTREME
RACING
37

RUSSIAN PAVILION AT THE 13TH VENICE BIENNALE OF ARCHITECTURE

At the 13th Venice Biennale of Architecture, Russia presented an exposition entitled "i-citi/i-land" dedicated to the Skolkovo Innovation City project. Commissioner of the Russian Pavilion Grigory Revzin, along with curator Sergey Tschoban (co-curators Sergey Kuznetsov and Valeria Kashirina), suggested an original concept bringing together the demonstration of architectural and town-planning projects with a presentation of scientific and technical innovations. The walls of main halls turned into surrealistic spaces are covered with metallic panels, which feature cut-out QR codes containing information about all stages of work on the Skolkovo project. Visitors to the pavilion obtain access to this information via tablet computers by scanning the codes and reviewing illustrative and textual materials related to them.

The second part of the exposition is related to the history of 37 classified Soviet science towns. The exposition is designed to accentuate the idea of secretiveness. The walls of pitch black halls have small holes in them behind which images from old chronicles and schematic drawings of these cities are located.

The two-part exposition demonstrates the interrelation and principal differences between science towns of the old times and the current project, declaring the concept of open urban environment, in which scientific achievements and high technologies become not an end in themselves but a means of improving the quality of human lives.

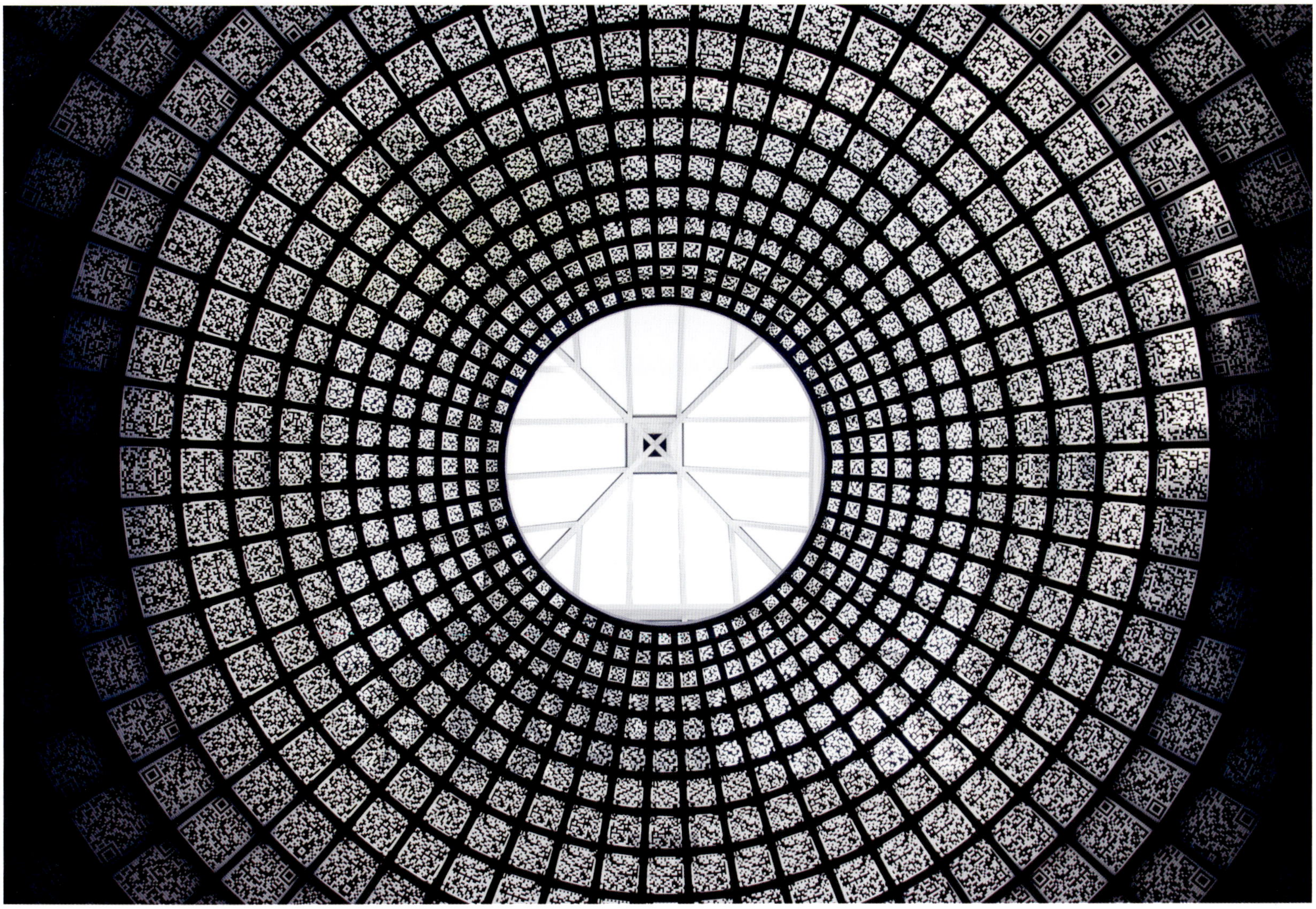

Architects
SPEECH Tchoban & Kuznetsov

Location
Russian

UNK project,
circle 10

SCIENTIFIC TOWNS AND CITIES
(Naukograds)
Akademgorodok by Novosibirsk
Apatity
Balashikha
Belozersky
Biysk
Chernogolovka
Dimitrovgrad
Dubna
Dzerzhinsk by Nyzhniy Novgorod
Dzerzhinsky by Moscow
Fryazino
Irkutsk
Koltsovo
Komsomolsk-on-Amur
Korolev
Krasnoobsk
Krasnoznamensk
Miass
Michurinsk
Mirny
Novouralsk (Zelenogorsk)
Obninsk
Obolensk
Protvino
Pushchino
Reutov
Sarov
Seversk
Sosnovy bor
Tomilino
Troitsk
Zarechny (Sverdlovsk Region)
Zelenograd
Zheleznogorsk
Zhukovsky
Znamensk (Kapustin Yar)
Zvezdny gorodok
i-land
During the cold-war period from 1945 until 1989 more than 60 gated towns and cities were created in Soviet Union for scientific and technological research. The existence of these cities was kept secret. They were everywhere in the country, and yet it was as if they did not exist. The people who worked within were isolated from society and were sometimes, for the sake of secrecy, given new names and surnames. These cities and their inhabitants were invisible except to the watchful eyes of the secret service.
For the exhibition we named this secret country 'i-land'. It is the subject of the exhibition on the ground floor of the Russian Pavilion.
The new city – the Skolkovo innovation centre – is an instrument for transforming science after the end of the cold war. This is an open city which is being created by some of the world's most acknowledged architects, and we hope to attract some of the world's advanced scientists. We called this city of the future 'i-city'. You will find it on the upper level.

Комсомольск
на-Амуре

Designer
Andrea Salvetti

"MAZZOLIN DI FIORI" & "NUVOLE DOMESTICHE"

"MAZZOLIN DI FIORI"

It is made up of over 2,000 anodised aluminium flowers in 4 colours, linked by pneumatic nails and a few screws.

Size D 550 cm, H 450 cm, 350 kg approx.

The large cupola or dome, structurally formed only by coloured alluminium flowers, appears light and transparent and provides an internal space suitable for moments of coming together, for reflection or rest. From within you can appreciate the natural shelter it provides, similar to that of a hedge for a wild animal, as well as the metaphysical comfort which looks to the future, an imaginary and creatively positive space; something evident in only a few contemporary art pieces today. Looking upwards, even the sky is improved. It is a large geodesic installation, a hymn to nature and its mysterious shape, both small and big, a synergy between architecture, sculpture and design in its expressive form.

"NUVOLE DOMESTICHE"

It is made of soldered aluminium springs, anodised and silver in colour.

Cloud 1: 215 cm x 120 cm x 86 cm, 105 kg approx

Cloud 2: 215 cm x 135 cm x 86 cm, 110 kg approx

Cloud 3: 2360 cm x 145 cm x 108 cm, 200 kg approx

Cloud 4: 108 cm x 120 cm x 68 cm, 40 kg approx

They are frothy silver shapes, indefinite and ethereal, more akin to the heavens than this world. These appear made of froth rather than water and they are firmly planted in a future scenario that carries everything along in its path.

They are tamed clouds in which to sit or lounge, to travel comfortably transported by wind and dreams. The clouds are made of alluminium coiled springs, where the metal is soft giving an impression of a metal never previously seen, that reacts with its own energy to living creatures. Hence, it will not be easy to tame them with one's body and understand their nature even if in principle the clouds have been tamed.

Domestic Nature

Design Agency
Kengo Kuma & Associates

Location
Niigata, Japan

Site Area
12,073.44 m²

Program
City hall, Assembly hall, Shops, Restaurants, Bank, Roofed plaza and Garage

Photographer
Erieta Attali

NAGAOKA CITY HALL "AORE"

With the growth of cities and their scale, public buildings of 20th Century were likely to be driven away to the suburbs, often as isolated concrete boxes in parking lots. The designers wanted to reverse this flow with Nagaoka Aore. They moved the city hall back to the center of the town and revived a real "heart of town" which is located in a walking distance from anywhere, working along with people's everyday life. This is exactly like the city hall historically nurtured in Europe, and embodies the idea of compact city in the environment-oriented age. They adopt the traditional method of "tataki" and "nakadoma" which is to function as a meeting point for the community, is no longer the mere concrete box – the space is gently surrounded by placid structure, finished with wood and solar panels.

6:48

6:58

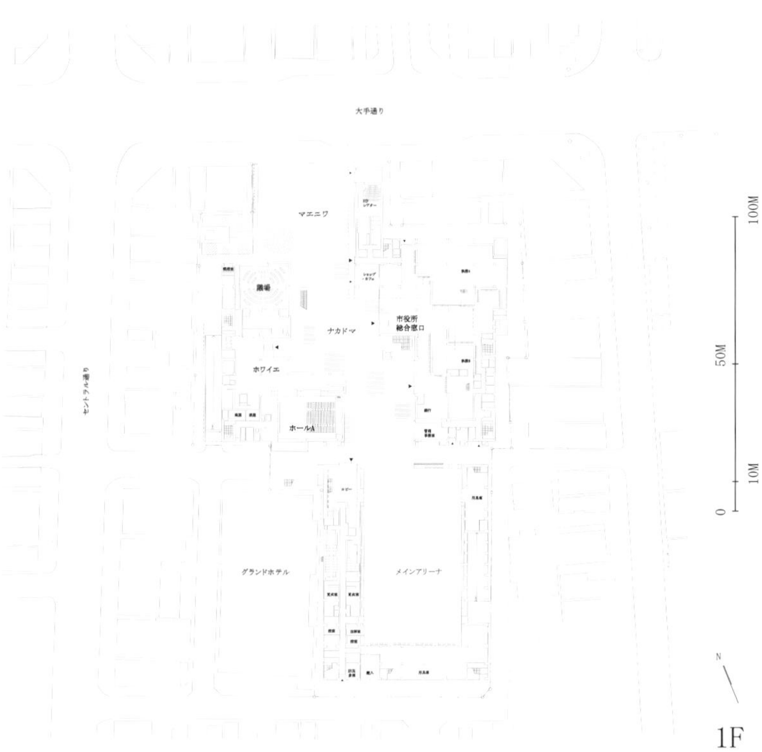

1F

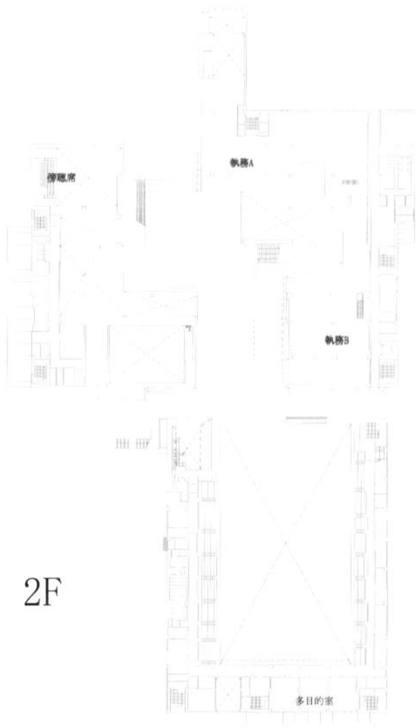

2F

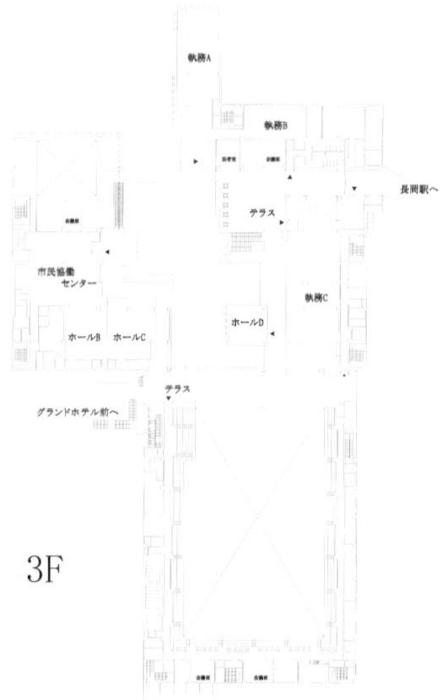

3F

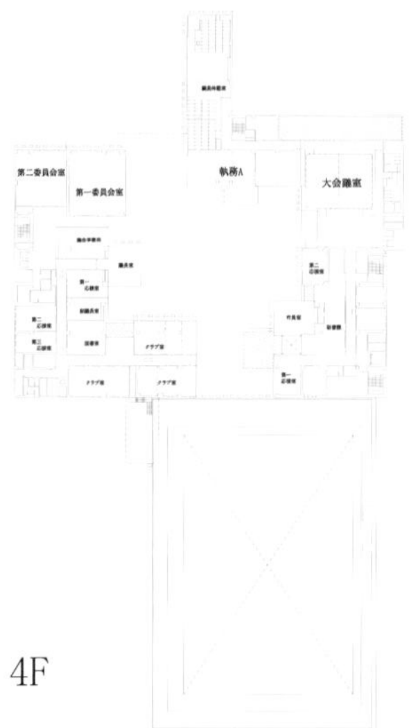

4F

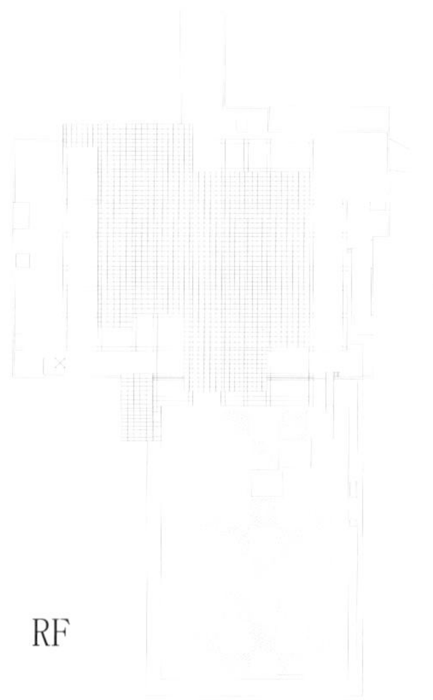

RF

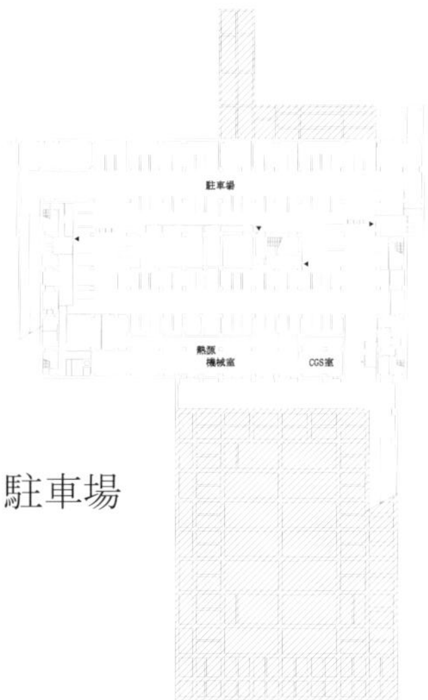

駐車場

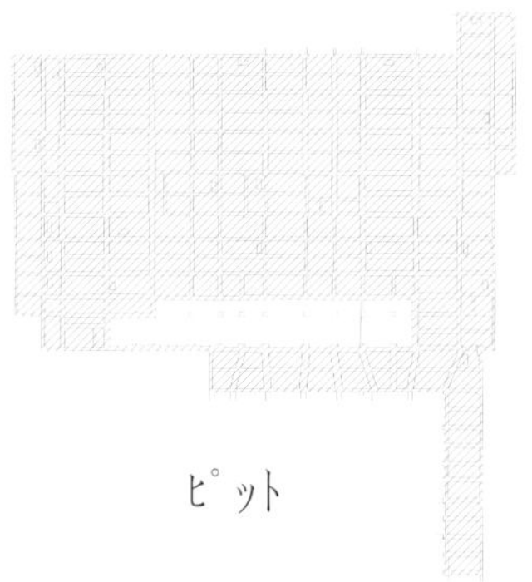

ピット

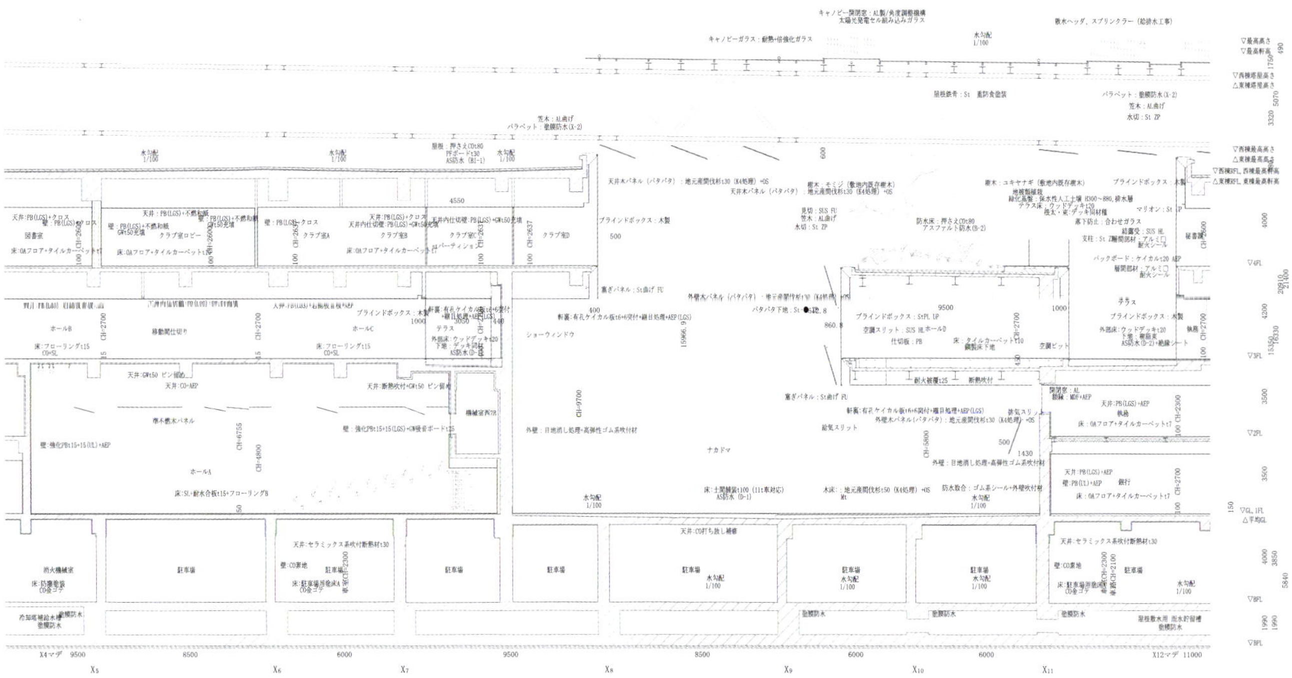

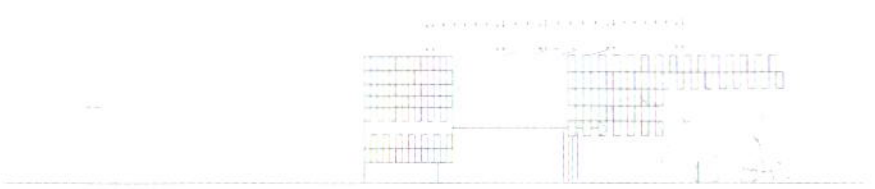

北立面図

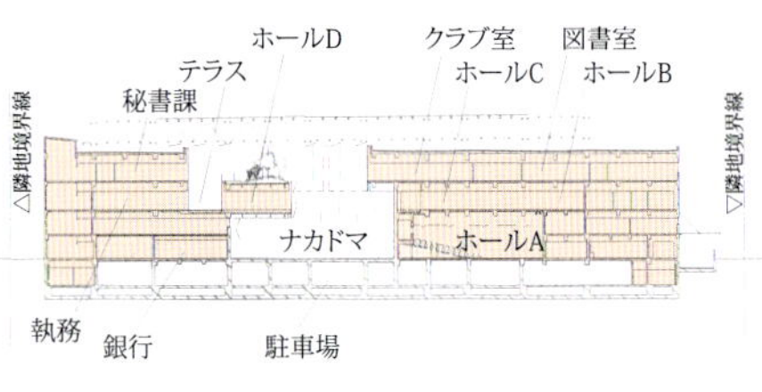

ナカドマ南立面図

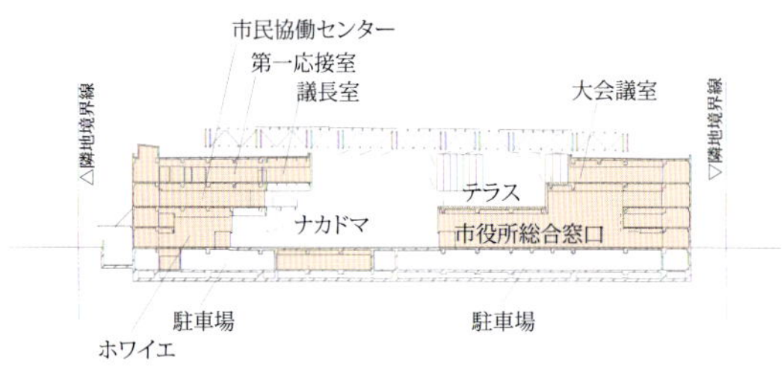

ナカドマ北立面図

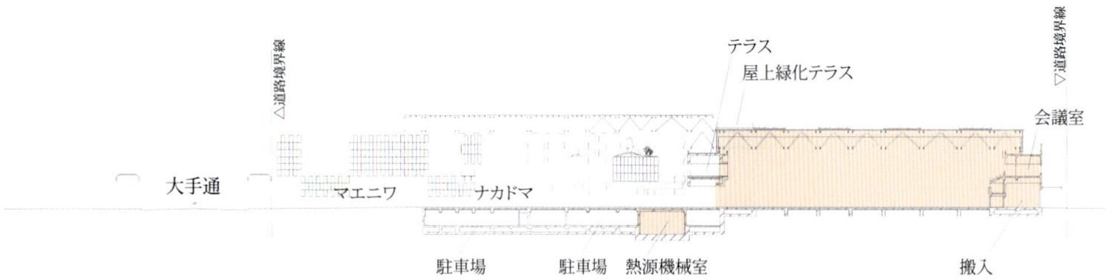

ナカドマ東立面図

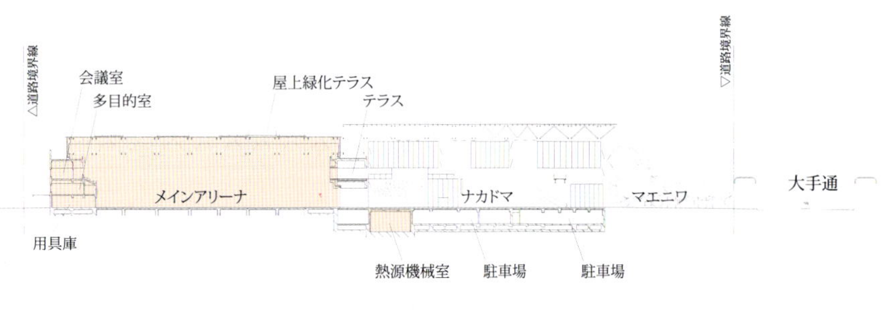

ナカドマ西立面図

THE OPPOSITE HOUSE

The Opposite House is situated within the large commercial development along the Sanlitun Street in the center of Beijing. The exterior is made entirely with glass curtain wall with green colored silk screen print with the modern interpretation of Chinese lattice screen pattern. The "green glass box" expresses itself as the urban forest in the vibrant street scene of Sanlitun at the same time acts as the veil for the guests staying at the hotel. The architectural of the hotel was developed from the traditional planning of Chinese courtyard house that encapsulates private quarters with a central courtyard. All the spaces in the hotel evolved around the large central atrium.

The seamless spatial sequence can be experienced through series of light screens made with different materials though out the public area to the guest rooms.

Design Agency
Kengo Kuma & Associates

Client
Swire Hotels

Location
Beijing, China

Site Area
4,548 m^2

Total Floor Area
14,328 m^2

Photographer
Michael Weber

PORSCHE PAVILION

The Dr. Ing. h. c. F. Porsche AG celebrated the opening of their Porsche Pavilion at the Autostadt in Wolfsburg in the presence of 200 guests of honor on June 12th, 2012. For the first time since its opening in 2000, the theme park receives another building structure in the form of the new Porsche Pavilion, which expresses the importance of Porsche within the Volkswagen Group family.

"The building is unique and its construction is extraordinary. This pavilion also has a symbolic and historical dimension, as it hints at the common roots through which Porsche and Volkswagen have been connected from the very beginning and will continue to be connected also in future", says Matthias Müller, CEO of Porsche AG. "As a worldwide leading automobile destination and communication platform for Volkswagen, we provide insights into our brands, values and philosophy for our guests. With the Porsche Pavilion we start a new chapter in the history of the Autostadt", adds Otto F. Wachs, Director of the Autostadt.

The organically shaped building is sitting — in mirrored location to the Volkswagen Pavilion at the central axis of the theme park and offers 400 m^2 of space for exhibitions and presentations. Its characteristic silhouette will become a distinctive icon amid the lagoon landscape of the Autostadt.

Curving lines and exciting bends make the Pavilion a dynamic yet reduced sculpture with its characteristics derived from the Porsche brand image. As designed by HENN, the structure captures the dynamic flow of driving with a seamless building skin. Its lines pick up speed and slow down just to plunge forward in large curves with ever-changing radii. A matte-finished stainless steel cladding forms the flush envelope of this vibrant structure, creating the impression of a homogeneous unity, whilst creating a continuously changing appearance depending on light and weather conditions. At the entrance the pavilion cantilevers 25m over the lagoon's water surface in front. Below the cantilever of the large asymmetrical roof, a sheltered external space opens up. This space is visually connected to the surrounding landscape, but forms

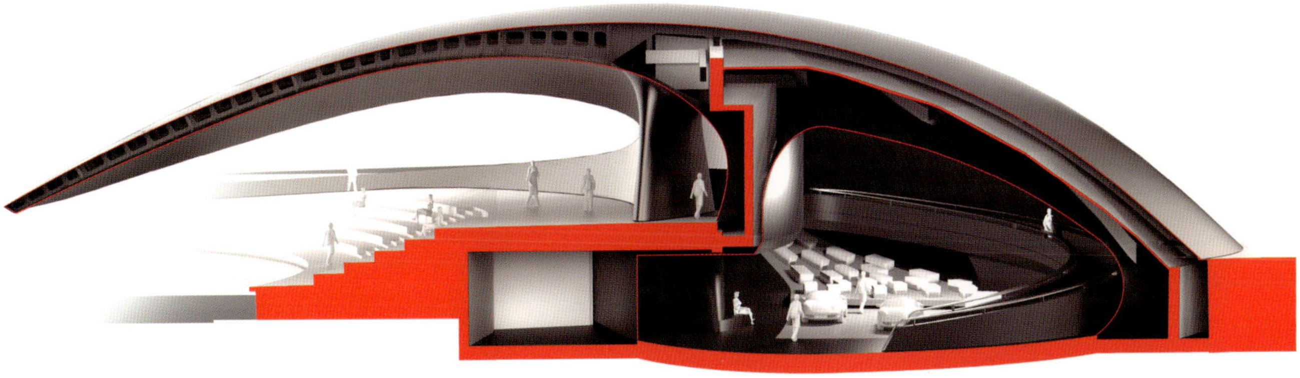

its own acoustic enclosure, providing seating for a few hundred guests. Architecture and landscape, interior and exterior as well as roof and facade are brought together by HENN in their architectural concept of a coherent, flowing continuum. The external area around the pavilion was designed by landscape architects WES and integrated into the overall concept of the theme park. This is how the new piazzetta creates a connection between the Porsche Pavilion and the adjacent Volkswagen Commercial Vehicles Pavilion by means of water features and trees. By walking around the sculptural Porsche Pavilion, further highlights of the Autostadt can be discovered.

Similar to the monocoque construction technology used for lightweight structures in the automotive and aerospace industries, the building envelope forms a spatial enclosure whilst at the same time acting as load-bearing structure. A total of 620 sheets of stainless steel cladding with welded ribs were prefabricated in a ship-yard in Stralsund and assembled on site.

Inside the pavilion a concentrated space opens up, allowing visitors to experience the sports car brand Porsche and its history, yet, casting aside the conventional limits and restraints to perception. The elliptically curved ramp embraces the dynamic principle of the architecture and leads the visitor to the lower exhibition stage areas. The exhibition and staging concept created by hg merz architekten museumsgestalter and jangled nerves combines evolution, engineering and the fascination of Porsche into an impressive image of future-oriented tradition. The Original Porsche – a 356 No.1 built in 1948 – is the starting point for a swarm of 25 silver coloured vehicle models at the scale of 1:3, on show in the main exhibition area.

Tradition and innovation, performance and day-to-day practicality, design and functionality, exclusiveness and social acceptance, these four antagonistic terms characterise Porsche's values and philosophy. They are also picked up as themes in short films. A film about the company history, sound stories about selected Porsche models as well as tablet PCs with further information about the exhibited vehicles make this visit's experience perfect.

Design Agency
HENN

Client
Dr. Ing. h.c. F. Porsche AG / Autostadt GmbH

Area
1,400 m²

Photographer
HG Esch

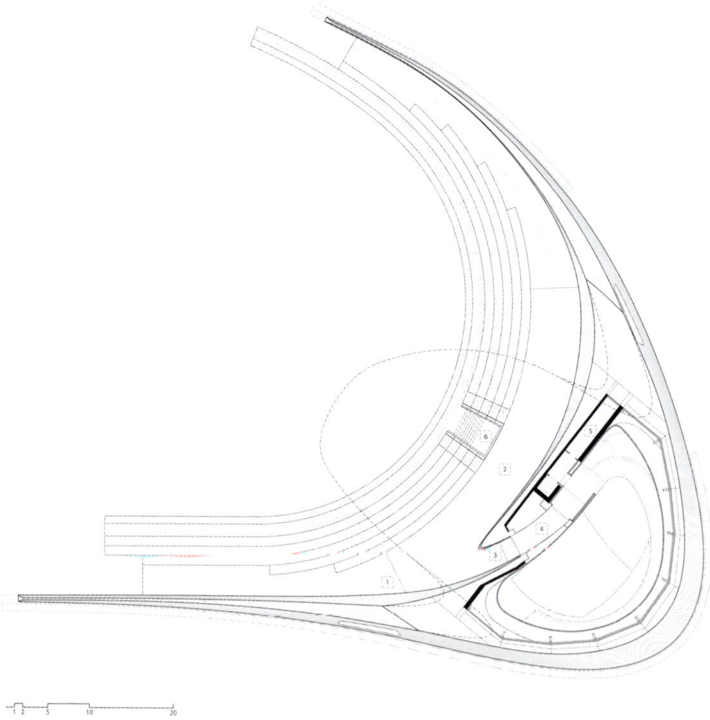

Erdgeschoss:
Eingangsebene

1. Rampe
2. überdachter Aussenraum
3. Eingang
4. Schleuse
5. Haustechnik
6. Ausgang

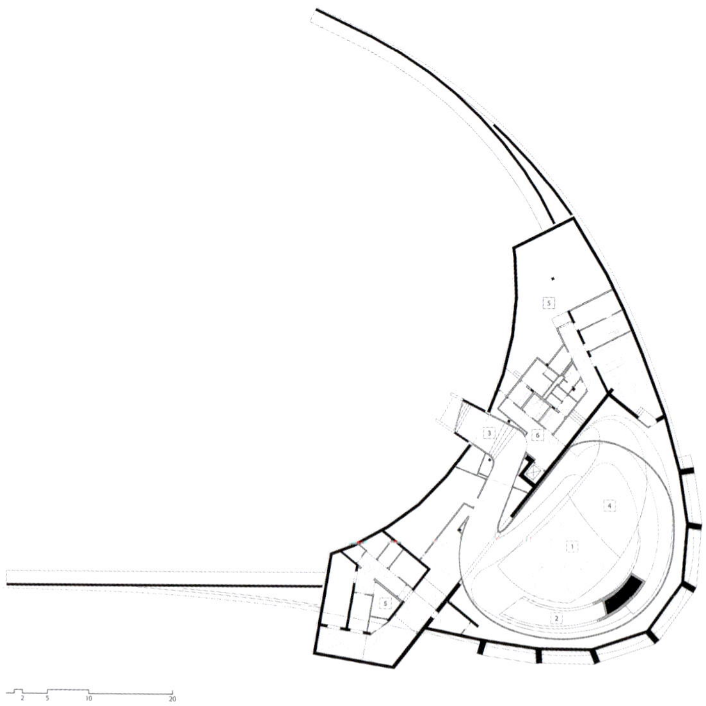

Untergeschoss:
Ausstellungsebene

1. Ausstellungsraum
2. Rampe
3. Ausgang
4. Inszenierungsfläche
5. Haustechnik
6. Sanitäranlagen

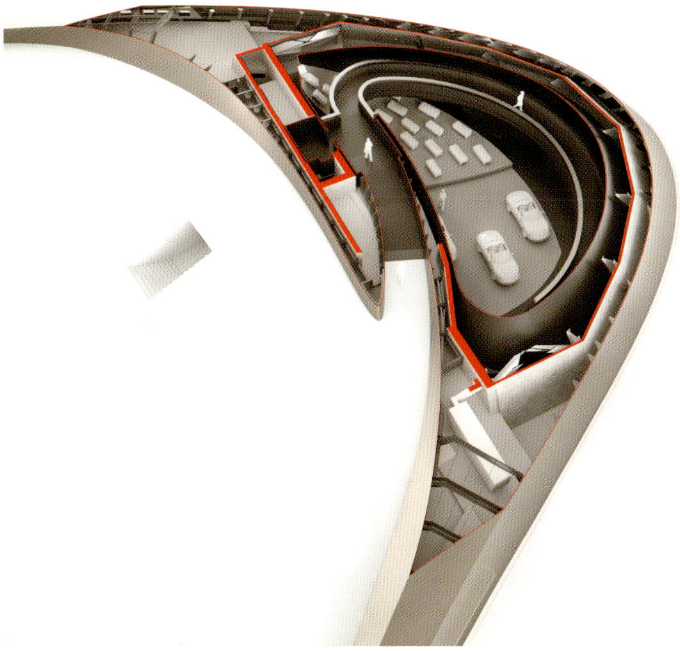

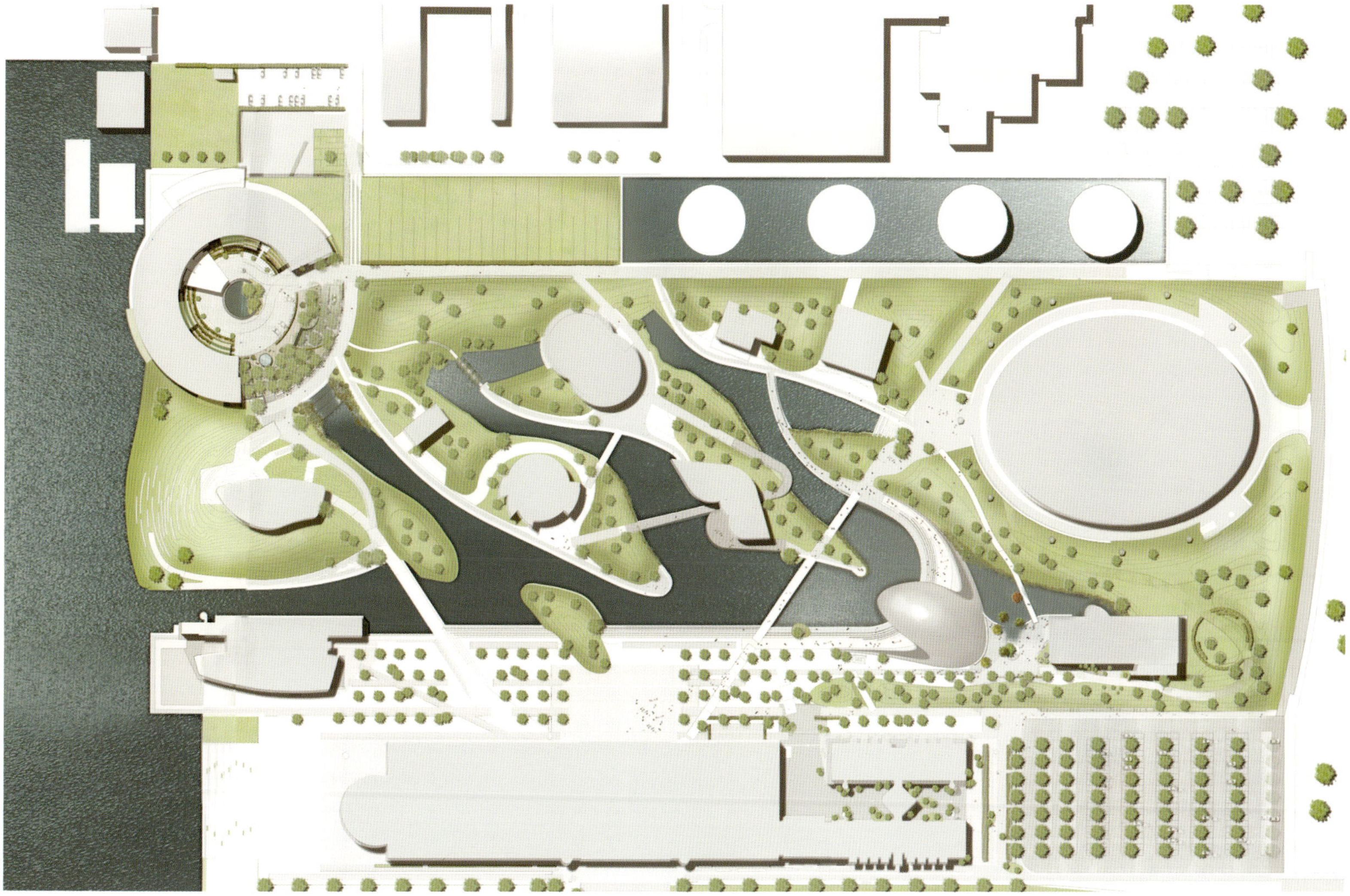

schaute ich mich um, konnte aber
von dem ich träumte, nicht finden.
selbst zu bauen."

Design Agency
heri&salli

Team
Lukas Allner, Monir Karimi

Structural Engineering
Bollinger-Grohmann-Schneider; Wien

Metallbau
Metallbau Fischer; Klagenfurt

Surface
SFK Tischler GmbH; Kirchham

Photographer
Paul Ott Photografiert

LANDSCAPE FENCE

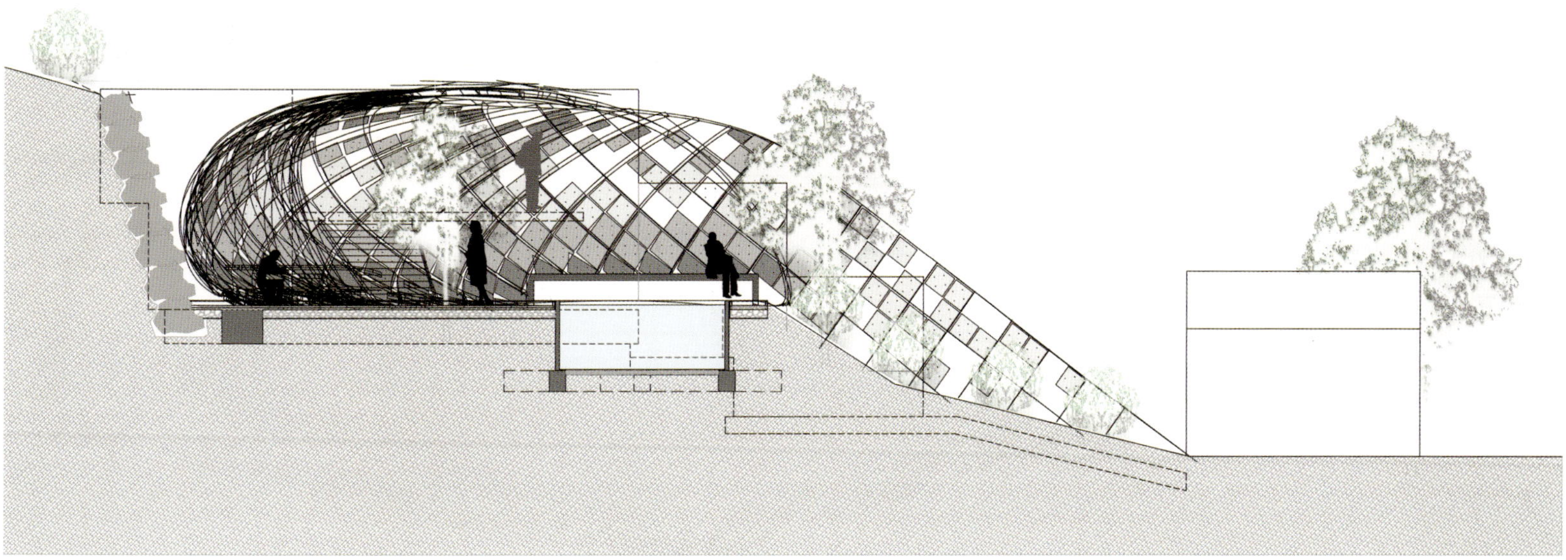

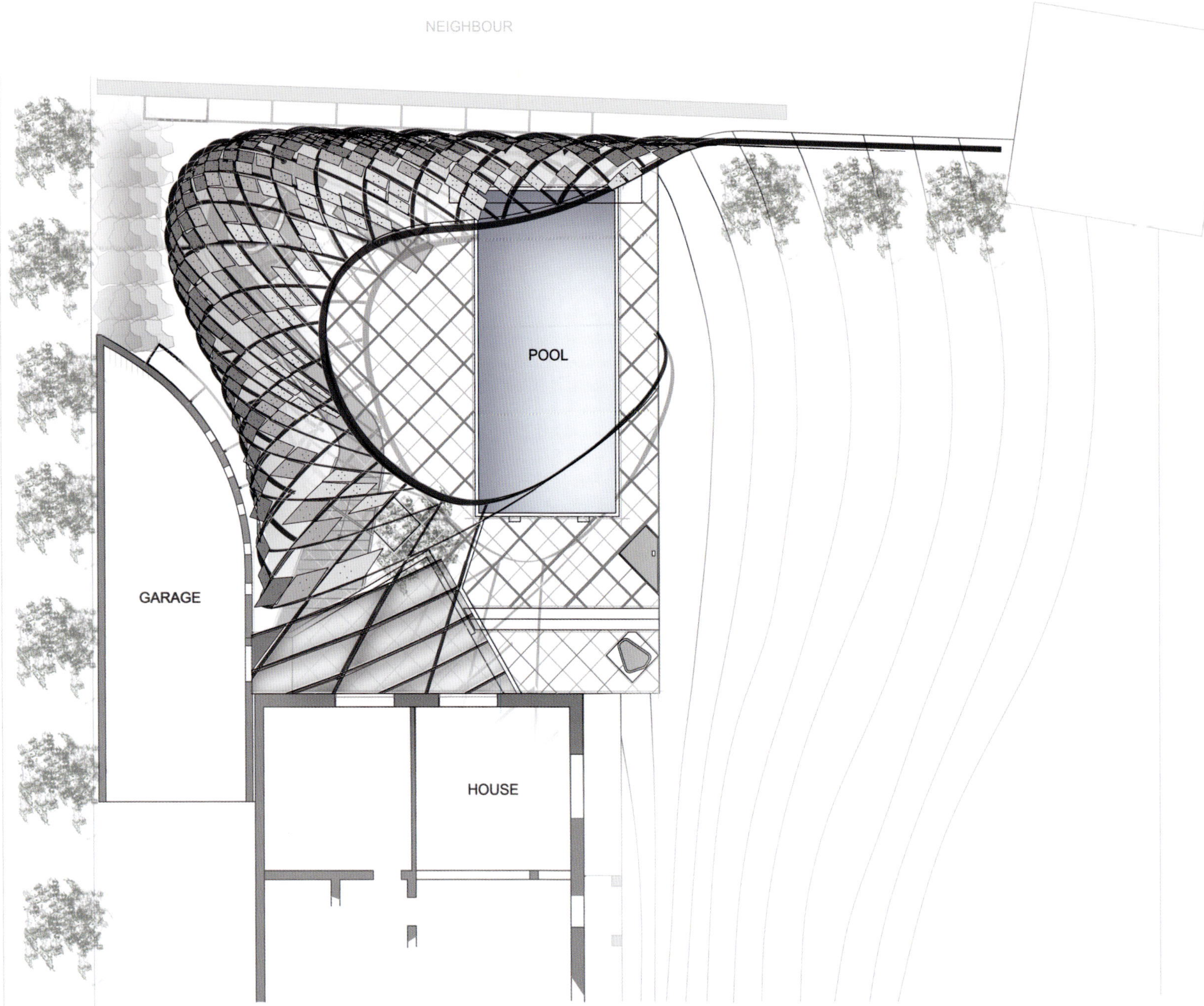
NEIGHBOUR
POOL
GARAGE
HOUSE

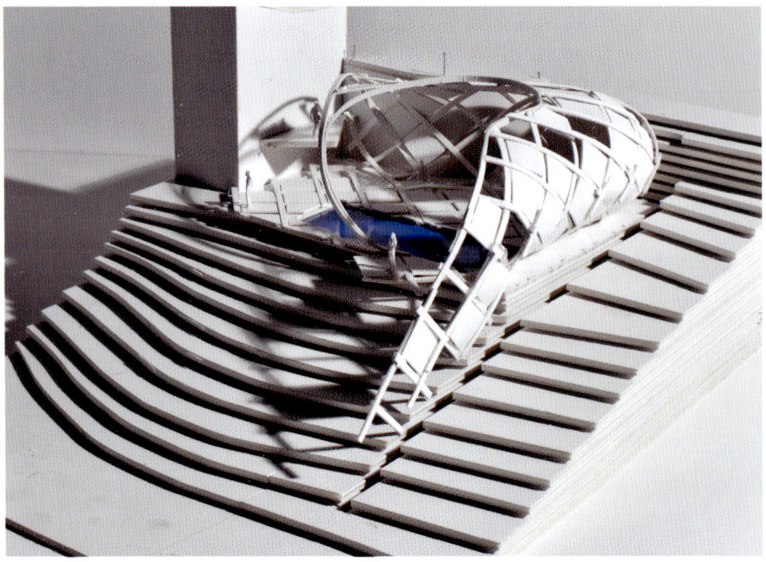

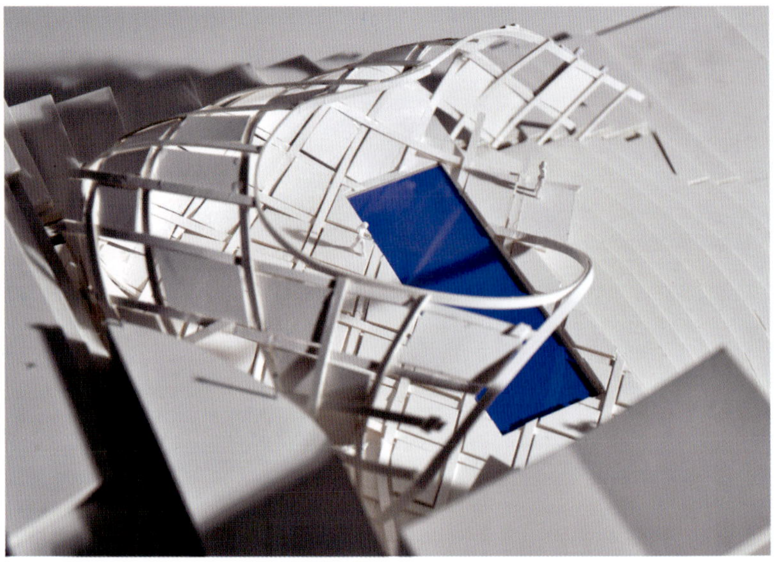

The architecture office heri&salli from Vienna conceived a steel structure similar to a cocoon round a swimming pool in the garden of a private builder-owner in Austria. With mounted panels and interior constructions which are more or less depending on their function the parametric organized spatial element describes possibilities of a usable and experience able surface.

Proceeding from the task to redefine an existing garden property with view of the lake, and simultaneously create provisions on views and a demarcation in direction of the surrounding properties and neighbors the theme of the classic rustic fence was taken up. In the simplest case a fence functions as protection or demarcation, a visualization of a line that wasn't visible before. In the further contest it serves as esthetical element or a representative sign and separates as a 2-dimensional element different areas. The designers formulate the fence based on different requirements as a 3-dimensional description of an existing garden. The fence itself becomes- preceding from a diagonal constructional arrangement therefore a possibility of space. With this in mind it doesn't demarcate the space, but creates it and renders it experience able, the function as a demarcation slides into the background and is only a by product.

The objective of the opening element similar to a cocoon is to create different spatial qualities and experience space. Partly covered, withdrawn and protected, then opening and finally in the middle or in the end in the water of the pool where you can swim out of it. The curves conveying a feeling of vastness make the space bigger than it is and create an optimal resonant behaviour inside of the house. Different integrated constructions like stairs, seats, lying areas or a table with backrest and pool covering are in its definition in a geometrical relation with the original construction; they emerge only to become part of the structure again. The integrated panels follow a dynamic course from the orthogonal edge into the described space, to develop in the central parts in relation to the steel structure from the inside to the outside or to dissolve more and more along the vertical. In this case architecture is an accumulation of possibilities in a described space and creates only the edges for a vast land in between.

The construction of the supporting structure can be described as an overhanging free concave form that is designed as frame construction with diagonally running circular tube profiles for out crossing and plate attachment. The frames consist of solid welded flat steel profiles. The not entirely closed shell is constructed with diamond shaped plates which are fixed by tabs on the diagonals and in case can be turned around their axis.

AKO BOOKS & TRAVEL

Tjep. was responsible for the interior design of the AKO flagship store at Schiphol Airport. For this project the designers were inspired entirely, and on nothing else but books, from the book displays, to the flying book light fixtures and the checkout counter reminiscent of a stack of books. And finally the book ribbon plays a central role in the entire store navigation.

The result is a spacious store, mainly filled with travel books, shop-in-shops by Lonely Planet and National Geographic and a reading area where you can read about places you're about to visit or dream away of exotic destinations that are high up on your bucket list.

Design team
Frank Tjepkema, Leonie Janssen

Builder
Hemi

Location
Schiphol Airport

Area
200 m^2

Photographer
Tjep: Yannic Alidarso and Martynika Bielawska

Top 10
Literatuur

AKO
Top 5
Non-fictie
English books
English books
Non-fictie
Non-fictie
Non-fictie
Reizen

AKO
Top 5
Literatuur
Spanning
Non-fictie
Reizen
Reisgidsen
Wenskaarten
Kinderboeken
lonely planet
TIP
MARTIN BRIL
IK HAAL JE OP
IK NEEM
JE MEE

VIRGIN ATLANTIC JFK CLUBHOUSE

Design Agency
Slade Architecture

Location
Queens, New York, USA

Photographer
Anton Stark / Slade Architecture TG

EXIT

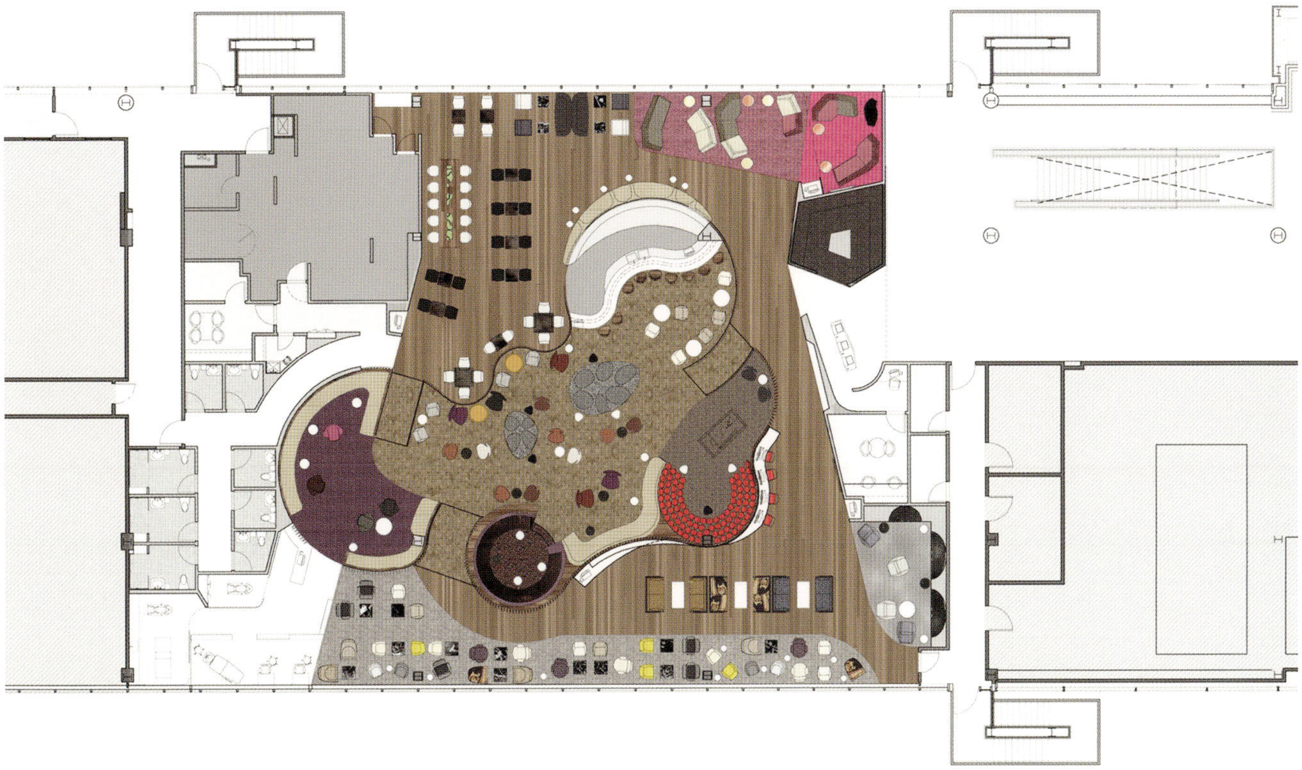

In designing the Virgin Atlantic Clubhouse at JFK, the designers sought to create a relaxed luxury space with a rich and distinctly "Uptown" Manhattan feel. The Clubhouse functions as a hybrid of private members club, boutique hotel lobby, restaurant and chic bar.

The 929 m^2 Clubhouse is bounded on two sides by full height expansive views over the jet ways, with Virgin Atlantic aircraft immediately below. With a direct view onto the iconic TWA terminal, the lounge picks up on this adjacency with subtle references to the glamour of the '60s air travel.

The overall strategy for the spatial organization was to create distinct areas that cater to different types of activities and different interactions between passengers. These areas are also organized by acoustical levels and temporal commitment. Acoustical zones: Quiet lounge, Talking lounge, cocktail lounge. Temporal zones: activities that require less time are closer to the entry, those that require more are furthest from the entry.

In the center of the clubhouse, the cloud shaped cocktail lounge is the central object that organizes the entire lounge. It defines the different zones; the interior of the cocktail lounge and the two spaces defined by the cocktail lounge on one side and the exterior windows on another, the talking lounge on the east side and the quite lounge on the west.

The cocktail lounge is enclosed by a diaphanous, curving "wall" of stainless steel rods and walnut fins that form a cloudlike shape in plan. This enclosure mediates views from and through the space and creates a series of distinct areas. The walnut fins are stepped to create and interior skyline. The designers designed two grey amorphous pebbles with soft asymmetrical seating niches and a fire-red ball sofa to provide unique seating landscapes. Fields of over 2,000 gold powder-coated spun cylinders hang from the ceiling creating a glowing, sculptural topography that dips into the space. This central area is the heart of the lounge, around and through which guests move in a rhythmic syncopated flow to the many unique Clubhouse amenities.

The talking lounge is furnished to encourage groups and interaction. The restaurant is located in this part of the clubhouse as well.

The quiet lounge has more individual seating. On one end, aluminum walls perforated with a pixelated cloud pattern are cutout to create "floating" seating pods. On the other end is Virgin Atlantic's first overseas spa and the first ever hair salon in any US business class lounge.

Smaller scale design elements reinforce the "Uptown" theme of the lounge. We designed two custom wallpapers that create a textured pattern from a distance. Upon closer examination they are composed NY icons: an optical composition using the Chrysler building and the Empire state building and a field of hot dog carts punctuated by red apples. In the bathrooms, white subway tiles and large-scaled black and white Sanborn maps of NYC landmarks complete the picture.

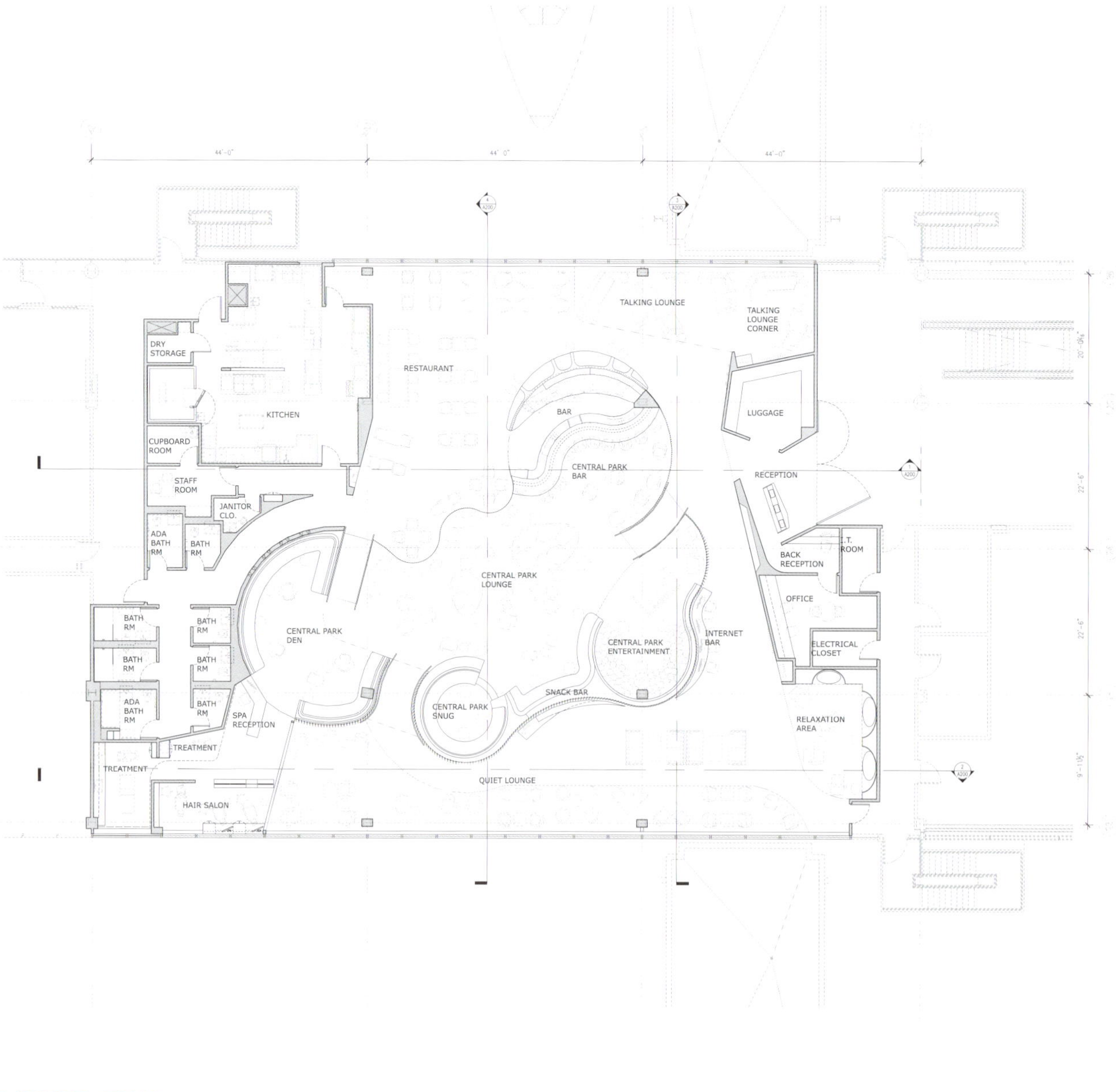
TALKING LOUNGE
TALKING LOUNGE CORNER
RESTAURANT
DRY STORAGE
KITCHEN
BAR
LUGGAGE
CUPBOARD ROOM
CENTRAL PARK BAR
RECEPTION
STAFF ROOM
JANITOR CLO.
ADA BATH RM
BATH RM
I.T. ROOM
BACK RECEPTION
CENTRAL PARK LOUNGE
OFFICE
BATH RM
BATH RM
CENTRAL PARK DEN
CENTRAL PARK ENTERTAINMENT
INTERNET BAR
BATH RM
BATH RM
ELECTRICAL CLOSET
ADA BATH RM
BATH RM
SNACK BAR
CENTRAL PARK SNUG
SPA RECEPTION
RELAXATION AREA
TREATMENT
TREATMENT
QUIET LOUNGE
HAIR SALON

EXIT
EXIT

SELLAND'S MARKET CAFÉ

Following the success of Ella Dining Room & Bar, voted "World's Most Innovative Restaurants"*, the Selland's Group once again commissioned UXUS to create a new signature Selland's Market-Café experience to be piloted at the Eldorado Hills Town Center development in California. The interior design and every detail of the customers' experience will embody Selland's unique philosophy of "high quality, hand crafted food". The Selland's brand can be described as serving up a perfect "slice of domesticity".

The mood at Selland's Market-Café will recall the simple comforts of a family country kitchen. Good food and friendly, personable service draw customers looking for a home-away-from-home. A romantic country kitchen inspires the décor. Touches of Americana recalling the brand's Sacramento roots are designed to feel like a personal collection of lovingly arranged items.

The color palette is comprised of shades of white, contrasted with natural wood and carefully placed touches of Selland's green. A mix of farm and café tables is arranged in a series of small "dining rooms" giving the front of house a relaxed ambiance. Casual elegance extends to small details, such as table settings and decorative lighting. Simple wood rolling racks and kitchen cupboards help breakup the restaurant into intimate dining areas, and display irresistible items for sale.

A blackboard menu frieze of picture frames creates the "family portrait" of meals on offer and is a distinctive feature of the space. Signature decorative lighting created from recycled jam jars and classic enamel kitchen lamps immediately evoke domestic bliss. A combination of mismatched white pressed-tin tiles wraps Selland's in a cozy patchwork, recalling turn-of-the-century kitchens and small-town delis.

An eclectic mix of chairs gives an air of informality and can fit up to 100 guests. Outside, ornate garden furniture in Selland's green contrasted with 50's inspired metal tables invites customers to spend some time to relax on the porch. Signature communal tables feature cast iron legs, casually assembled to look like homemade creations.

This concept is to be rolled out to other locations at the beginning of 2013.

Voted "World's Most Innovative Restaurant Interiors" by Fast Company's Co. Design

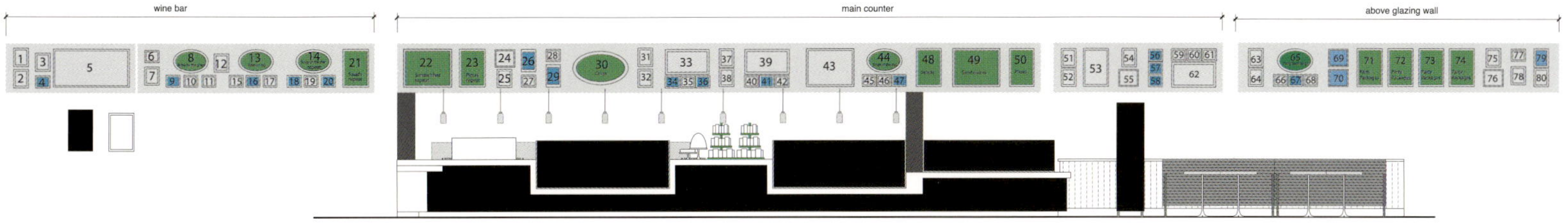

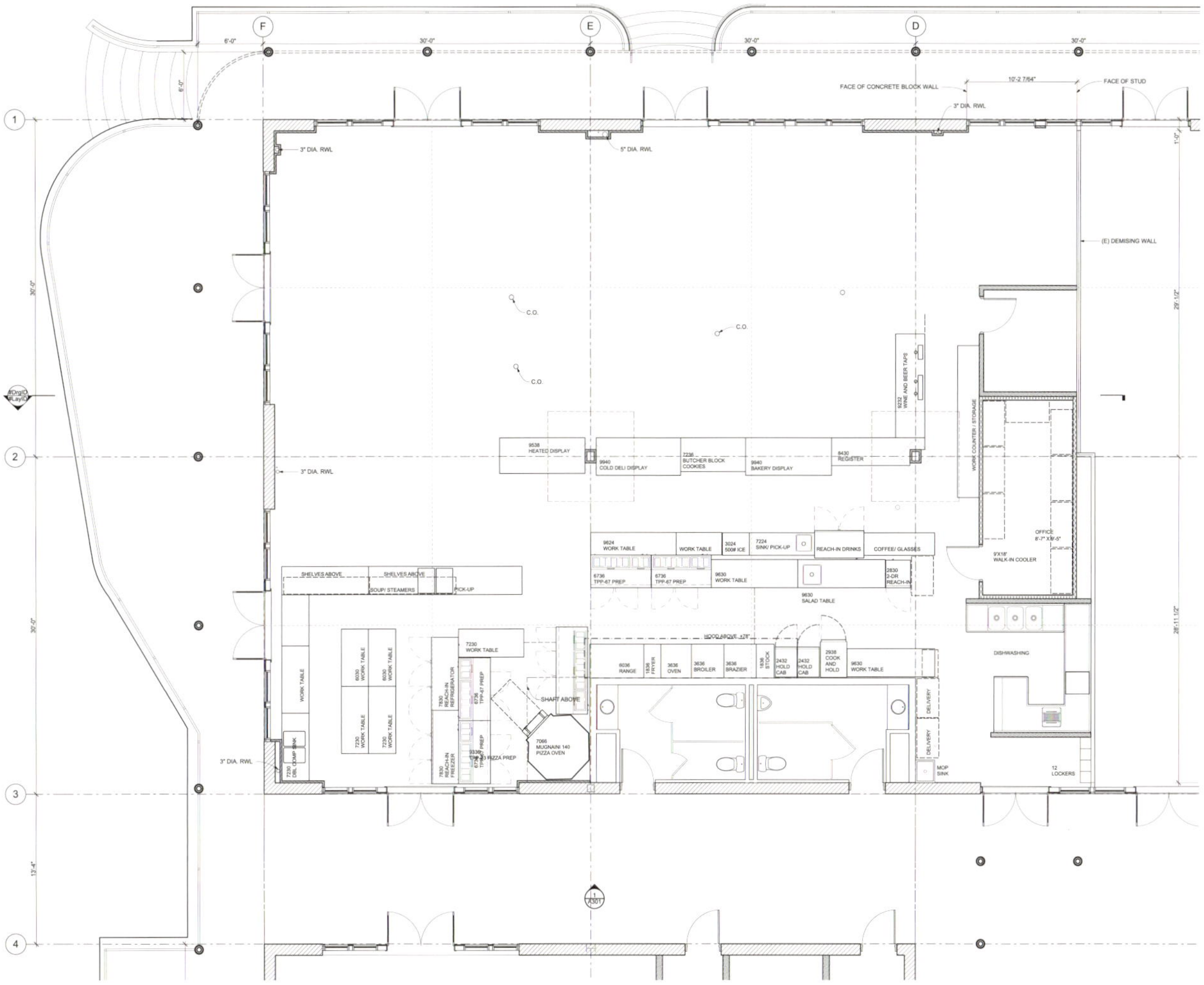

Design Agency
UXUS

Architect
Louis Kaufman

Contractor
Unger Construction

Client
Selland Family Restaurant

Location
El Dorado Hills Town Center

Area
176 m^2

Photographer
Donahue Photography

Beer on Tap

Beer on Tap
Wine By the Glass
Eat Dessert!

15% discount
on all cases!

Retail Wine
Buy a bottle here:
No Corkage!

all you
need is
LOVE

Design Agency
Stone Designs

Client
Maxibread

Project Team
Eva Prego & Cutu Mazuelos

Location
Saint Petersburg, Russia

MAXIBREAD

Maxibread is a bakery with a café area and its concept arises from the merge between Russian and Western European culture. This fusion is what we can call the new style of Russian design.

The main idea was to emphasize the values of quality and warmth, so the designers managed to create a cosy place using decorative elements inspired by the Russian tradition and combined with ambient lighting which was a key factor to enhance and create small spaces.

The bakery area is demarcated by a carved wooden roof inspired by the traditional Russian houses. For the counter the designers decided that the marble would provide the quality, weight and elegance necessary for the project. The café area was intended to contrast with the bakery area. It recreates a comfortable environment with a very faint and pleasant light.

THE WAHACA SOUTHBANK EXPERIMENT

The Wahaca Southbank Experiment is a two-storey temporary restaurant installation, constructed from eight recycled shipping containers that have been "washed up" on to the outdoor terrace of the Southbank Centre on the River Thames in London.

The idea for using the shipping containers was developed to serve as both a reminder of the working history of this part of the river, and for more practical reasons as their limited height meant that the design team was able to insert two floors into the height of a single storey space.

Situated against a heavy concrete backdrop, each container is painted in one of the four vibrant colours ranging from deep turquoise to straw yellow. One of the top floor containers has been cantilevered out over the restaurants entrance to create a canopy above the ground floor. On the upper level, the effect of this cantilevering heightens views from the upstairs bar out over the river.

Inside the restaurant the front and back containers are connected via a glazed link, which not only houses the stairway connecting the two floors, but also helps to flood the space with natural light. Each of the containers has then been given its own character with a mix of bespoke, new and reclaimed furniture along with distinct lighting designs.

Outside, there is a wide variety of areas in which to sit, from the booth seats, built in to the raised timber deck around the building, to the first floor terrace bar, to the street bar overlooking Queen's Walk.

Design Agency
Softroom Architects

Client
Wahaca Group

Location
Southbank Centre, London, UK

Structural Engineer
Price and Myers

M&E Engineer
TR Mechanical Services Ltd.

Principal Contractor
du Boulay

Lighting Design
Kate Wilkins

Project Manager
Bright Spark Ltd. for and on behalf of Wahaca Group

wahaca

CHI.LL.I.
/PPR
350,000 SHU
9/10
MASA.
FLR
TACO
36.801 KG
86.161 LBS
A.V.O
CADO
9000

Design Agency
Stone Designs

Project Team
Eva Prego & Cutu Mazuelos

Location
Madrid, Spain

FARMACIA DE LOS AUSTRIAS

The Farmacia de los Austrias (De los Austrias Chemist) is placed in one of the most emblematic areas of historical Madrid.

Designers' initial idea was to create a new space typology, in which tradition and vanguard merge in such a subtle way that originate a slow and deliberate dialogue in which no element stands out of the rest, creating an almost musical harmony.

Products are displayed in really thin metallic structures standing in the bluish walls, creating a sweet and warm chromatic range that makes us feel at ease. This space transmits that they are attended by real professionals, but with a more human touch than usual.

Some details such as the white marble counter help to strengthen the concept of the "well done job" that oozes the old; while other materials like the tiled floor, embrace us in a warm and close atmosphere.

It is a project in which, due to its nature and small size, even the slightest detail has been taken care of, creating an enormous sensory universe that makes the visitor enjoy a most gratifying experience.

Eucerin
garydol
garydol
cariax
cariax
KIN
KIN
SENSODYNE
Lacer
Lacer
Oraldine
Oraldine
Oraldine
LISTERINE

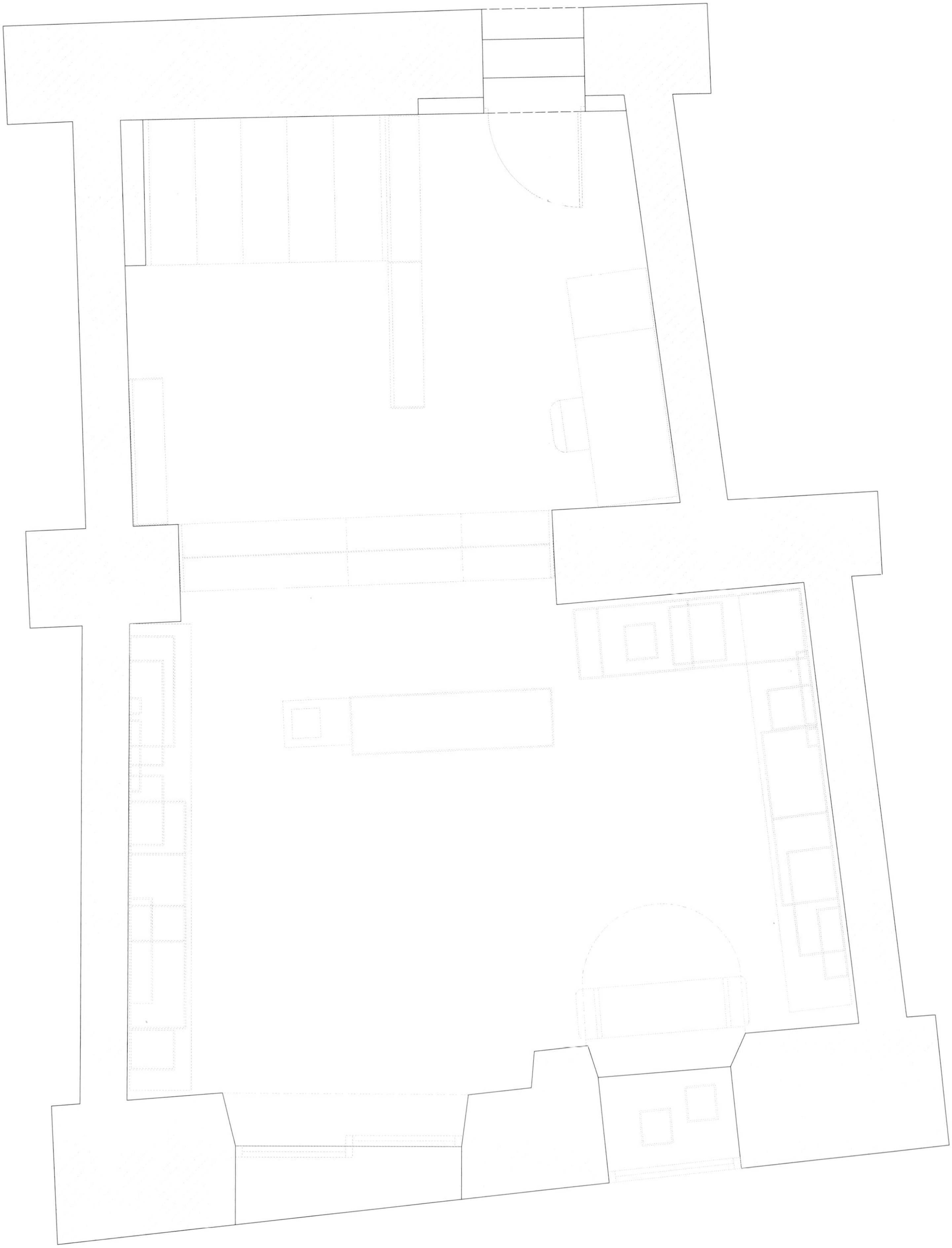

Ibercal-D
Nutriben

Pharmaton
Ibercal-D

YOGOOD

Architect
Marcos Samaniego Reimundez

Design Agency
MAS · Arquitectura

Photographer
Ana Samaniego

When designers begin an intervention, they must consider not only a solution having the premises. They must also focus on the whole building to obtain the best outcome for those involved (customer, citizens and technical equipment).

Therefore, with this project, they recover the identity of the building to pose the detailing of the woodwork, following the existing line on the upper floors. Essence that had lost the shop and that detracted from the overall image of the building.

In the premises, they have managed to harness the full width between party wall, thereby achieving greater breadth to link the local and building access on one front.

The designers designed the counter as the main element, so that the whole space could work around it. Thus they create a fully functional space through an element that plays with the height to create different spaces. On the lowest level, it creates a waiting area with a wood finish for being comfortable. At the intermediate level, it creates an exhibition space, which allows us to appreciate the product. Finally, the highest point, more functional, it allows complete the design of the furniture giving a pleasant and comfortable interior space.

The counter is completed with an upper glass, all at the same level, which gives the feeling of cleanliness, transparency and quality.

The corporate graphic design element, by waves, is used on the counter in the form of low relief. The corporate color is applied only in the lamps that illuminate the counter and in the graphic design.

It has used the space of the counter and won to the party wall for extending the work area and hiding a small sink.

The end of the counter opens diagonally to the warehouse, incorporating the ice cream machine, and the shakes one in the background.

All openings in the interior of the premises have been made with blunt corners, in line with the local graphic design.

MAS · ARQUITECTURA
www.mas.es
yogood
Yogo, el helado de yogurt
superbueno

yogood
Limón
Naranja
Fresa
Manzana verde
Sandía
ice-fruit 2,70€
maxiyogo 3,50€
TropiGood
Yogo natural
Fresas
Plátano
ChocoGood
Yogo de chocolate
Leche
Plátano
yogo 2,70
+ 2 toppings
RobinGood
Piña
yogodrink 3,50€
FreshGood
Yogo natural
Ice-Fruit de naranja
Kiwi
miniyogo 2,30
+ 1 topping
Yogo, el helado de yogurt
superbueno

Limón
Naranja
Fresa
Manzana verde
Sandia
ice-fruit 2,70€
maxiyogo 3,50€
yogo 2,70
miniyogo 2,30
TropiGood
ChocoGood
RobinGood
FreshGood
yogodrink 3,50€
Yogo, el helado de yogurt
superbueno

Fideos chocolate
Chocolate
MiniNubes
Yogo, el helado de yogurt
superbueno

HOUSE OF FLAGS

Situated on Parliament Square, House of Flags pays tribute to the diversity of all the nations taking part in the 2012 Olympics and Paralympics. 206 panels depict the flag icons of the nations involved in the Games and combined make up a large building jigsaw: a united "house" of symbols, shimmering colors, shadows and perforations, that invites everyone to experience a matrix image of the world as well as a portrait of multi-ethnic London. House of Flags creates an engaging backdrop for thousands of visitors who explore the installation with excitement until they find their flag and are proudly photographed in front of it.

The installation, designed by London-based AY Architects, was commissioned by the Greater London Authority after an invited international competition for the Mayor of London's WONDER series of "Incredible Installations".

The exterior of the installtion demonstratively produces a global image made up of the flag iconography, while the interior is defined by an abstracted interpretation of the flags, creating a more unified experience. The design acknowledges the World Heritage setting and highly political status of the square, historically charged by protests and demonstrations. Countless stories can be told about the symbolic significance of both the flags and the site and the paradoxical issues of identity and security that emerged in the dialogues between the architects and the various stakeholders during the design process. House of Flags collates politics, graphics and architecture into one gesture.

The installation is a free standing structure measuring 17 m long x 8 m wide x 4.5 m high. It is made of 206 FSC certified birch plywood panels and over 400 laminated connection components, of which there are 8 different types. It stands on 42 pre-cast concrete foundational blocks.

The panels are CNC cut and the majority of them have cut-outs of symbols and perforations. The top panels are more perforated and lighter whereas the bottom ones are more solid and therefore heavier. Panels are stacked like a house of cards with alternating orientation from row to row. The result is a complex layering of colour, light and reflection seen against the imposing backdrop of the surrounding historic buildings.

Design Agency
AY Architects

Structural Engineer
Price & Myers

Fabrication
Grymsdyke Farm

Printing
Signet Signs

Installation
Bolt & Heeks Ltd.

Location
Parliament Square, London, UK

Photographer
Nick Kane

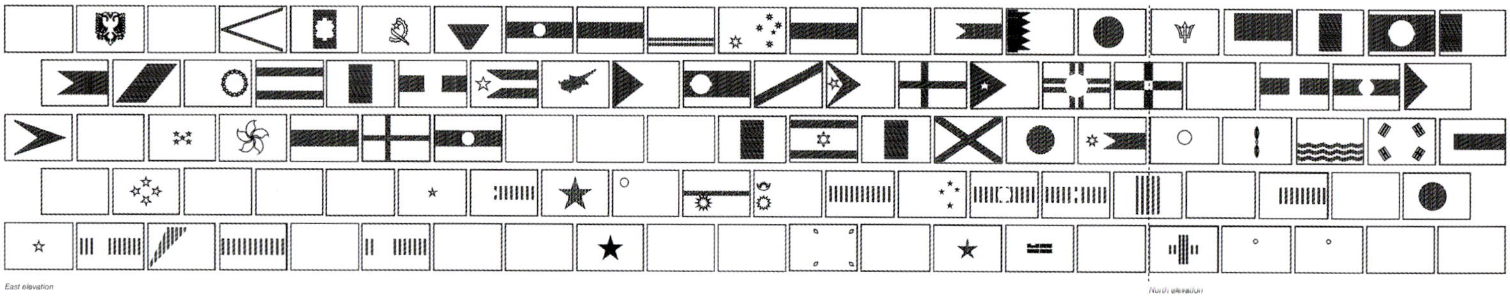

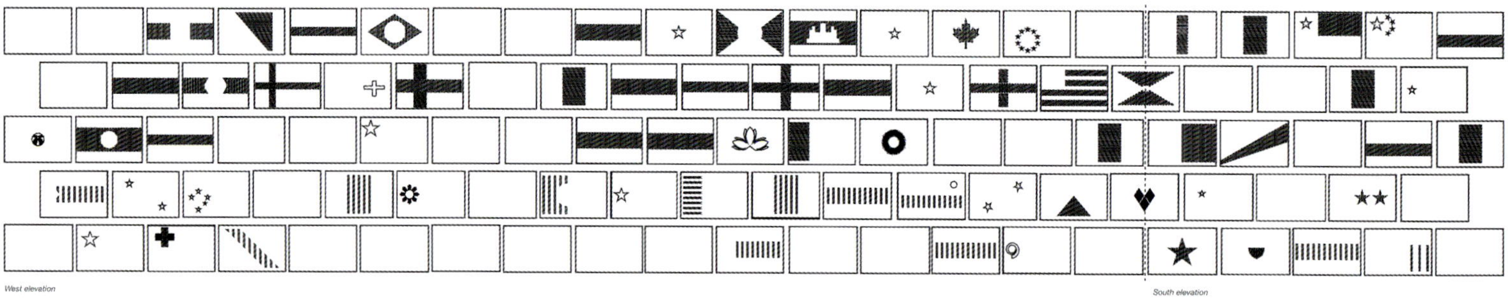

The graphic of each flag was printed directly on the plywood panels using a large format UV flat bed printer, which resulted in a crisp image read through the timber grain surface. The back of each flag panel was left to show the natural material finish. Curiously, the untreated plywood finish of the internal elevations plays with the weathered limestone of the Houses of Parliament backdrop. At the same time the vibrant colors of the external elevations suggest an inversion of the exuberant colours of the interior of the Houses of Parliament.

During the day the structure works as a shadow modulator with the shadows of its perforations shifting from east to west. When the sunlight is sharp soft layers of colour light, produced by the vibrancy of the colour-printed panels, are reflected on the natural plywood panels next to them. At night the structure is lit from within, glowing as an inhabited "house" and showing the emblem cut-outs appearing as silhouetted figures.

The structure is flat-pack, demountable and entirely reconfigurable. As a kind of large three-dimensional puzzle it can be reinstalled in new configurations and flag hierarchies of various shapes and sizes.

Its current composition presents the flags in alphabetical order, never touching or intersecting, and carefully considered so that certain cut-outs are not offensive or seen from the back. AY Architects worked with The Flag Institute, the world's leading research and documentation centre for flag information, to determine how the design of the panels and their connections could respect these protocols.

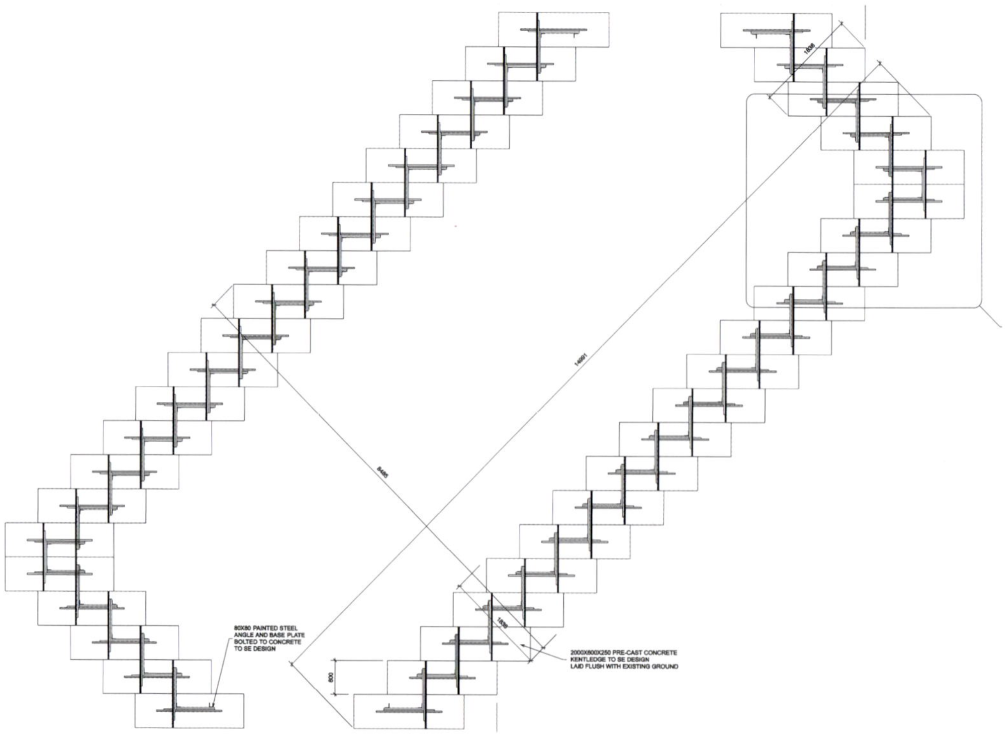

لا إله إلا الله محمد رسول الله

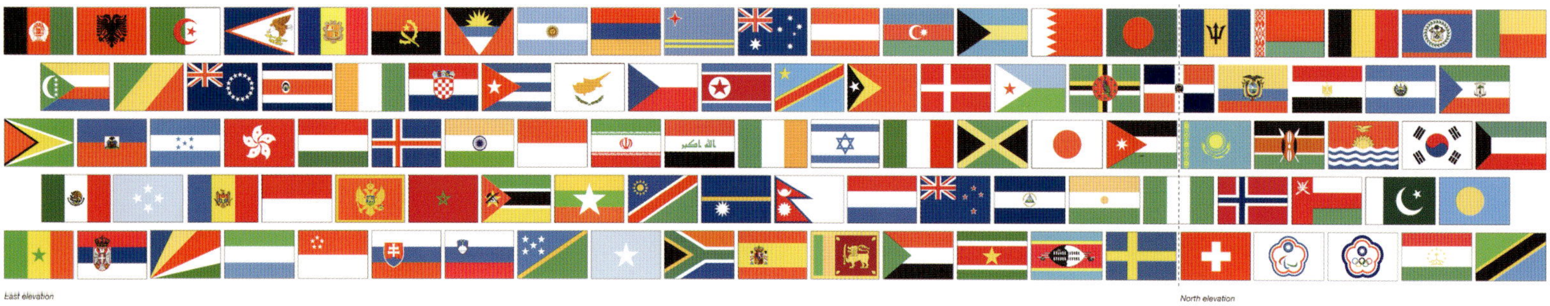
East elevation
North elevation

West elevation
South elevation

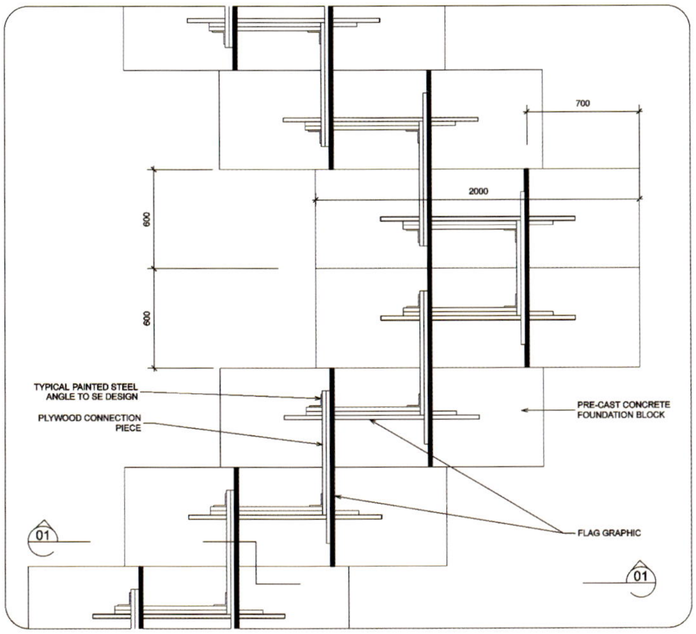
700
2000
600
600
TYPICAL PAINTED STEEL ANGLE TO SE DESIGN
PLYWOOD CONNECTION PIECE
PRE-CAST CONCRETE FOUNDATION BLOCK
FLAG GRAPHIC
01
01

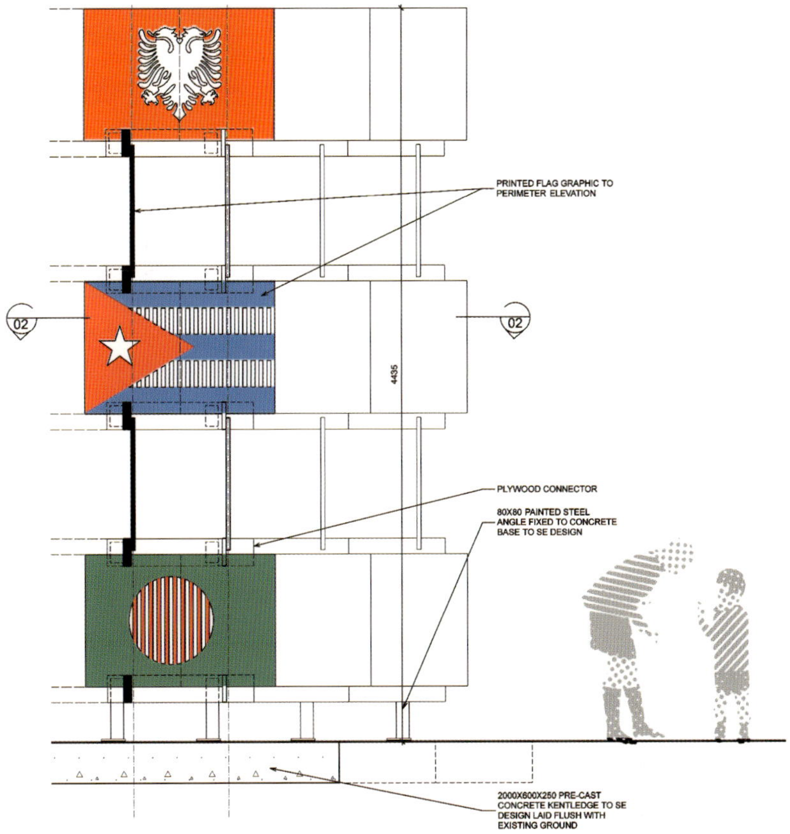
PRINTED FLAG GRAPHIC TO PERIMETER ELEVATION
02
02
4435
PLYWOOD CONNECTOR
80X80 PAINTED STEEL ANGLE FIXED TO CONCRETE BASE TO SE DESIGN
2000X600X250 PRE-CAST CONCRETE KENTLEDGE TO SE DESIGN LAID FLUSH WITH EXISTING GROUND

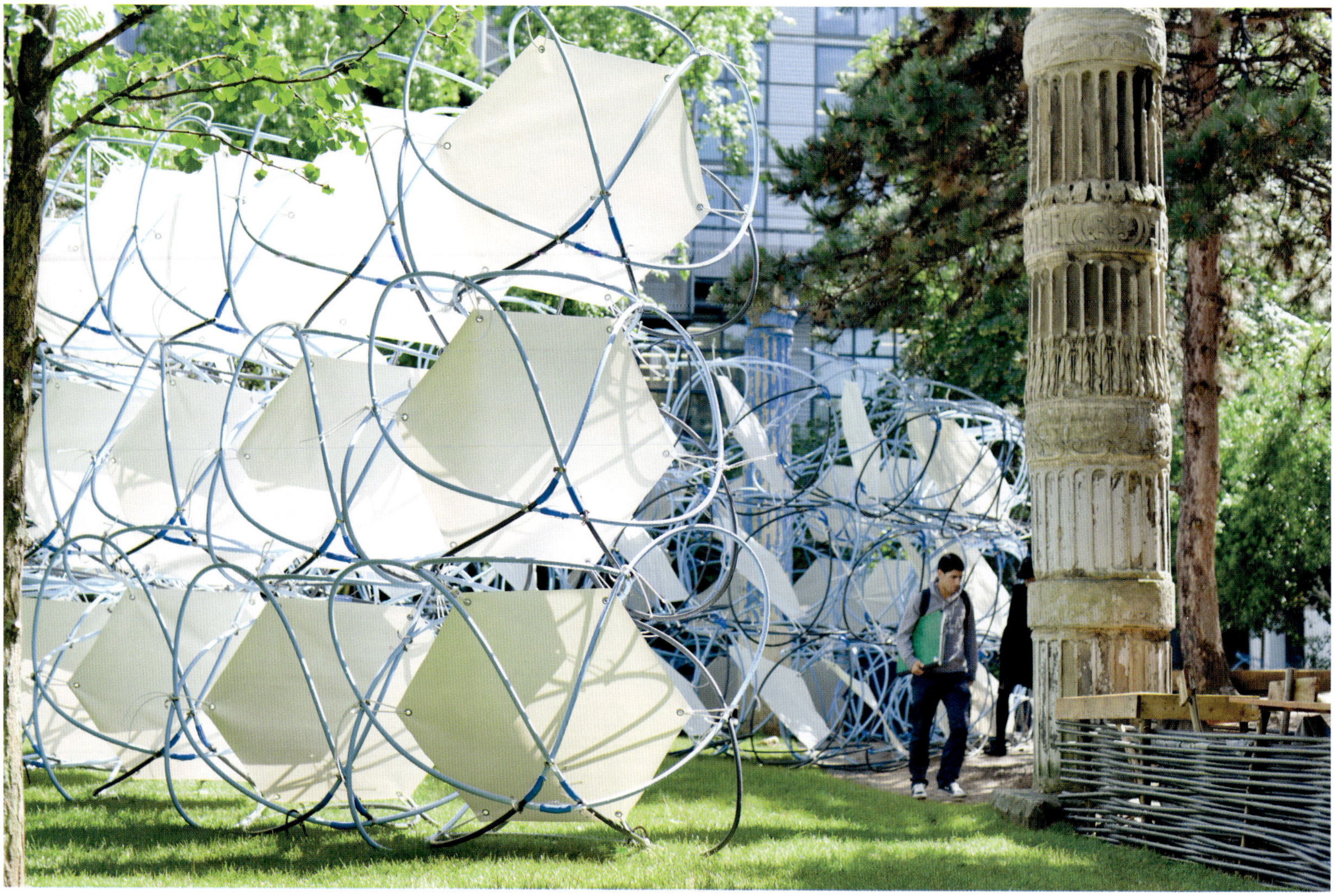

Architect
Ball-Nogues Studio

Photographer
Arnaud de Prez

THE PAVILLON SPÉCIALE

The Pavillon Spéciale is an installation designed and built by students of the Ecole Spéciale d'Architecture under the direction of Ball-Nogues Studio. The installation can be arched and curled at full scale to form different types of space befitting the university's summer program. The installation creates a sense of place while providing a respite from the sun and rain.

The pavilion is a unique structure. In architecture terminology, the phrase that describes a system whose form is derived from the deformation of its materials under force is "form active". This type of structure is difficult to study using software. It often requires architects to explore their designs by testing full-scale mock-ups, and using that empirical information to help inform the process of digital modeling, which is studied in the studio rather than in the field. Students engaged in this iterative design process with Ball-Nogues.

The structure is comprised of approximately 200 "cells", each made from locally sourced plastic tubing bent and curled in custom jigs designed and constructed by students. To provide shade, each cell has locally sourced fabric membrane spanning between the tubes. The cell module is a very effective way of constructing a temporary structure: each can be transported as a flat unit and rapidly assembled on site; when it is time for the structure to come down, dismantling and transportation to a new site is easy.

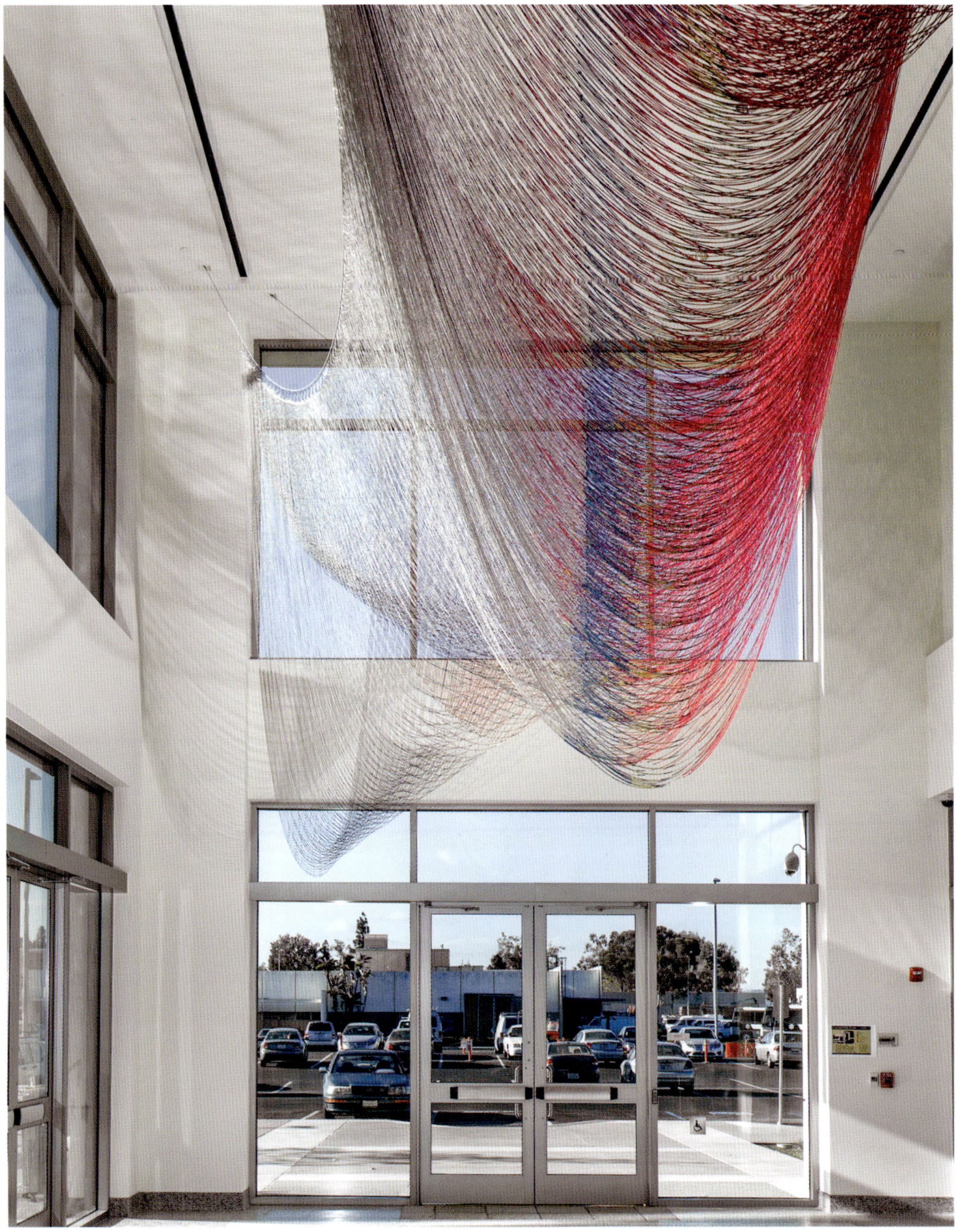

Architect
Ball-Nogues Studio

Photographer
Scott Mayoral

WATERLINE

Waterline resembles a thickened atmosphere of ghostly waves within the double high entryway of Building 204. It is neither solid nor emptiness but has qualities of both. Seventeen thousand segments of painted stainless steel ball chain, totaling over 10 miles in length make up this work. By integrating digital computation with hand production techniques, Ball-Nogues meticulously combined the segments to form an array of "catenaries" that span the ceiling. In mathematics, a catenary is the shape of a curve formed by a chain hanging between two points.

Composed of seven colors, the chains make an intricate system of overlapping curves. The result suggests a three-dimensional abstract painting that looks differently depending on one's vantage point. From one angle, the viewer sees hard-edged geometric shapes in distinct color; from another angle, the same colors blurred to make a vapor-like composition.

In naval engineering, the term "waterline" refers to the contour made by the hull of a ship meeting water. This Ball-Nogues installation includes a field of magenta color that is parallel to the ground plane catenaries. Analogous to a waterline, this feature becomes reference for gauging the discrepancies between the "theoretical" models generated within the computer and the physical reality of the installation constructed from the data output by the computer.

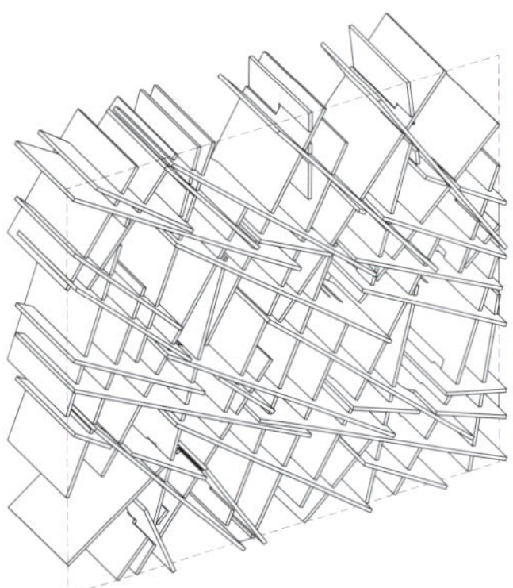

Egg-Crate

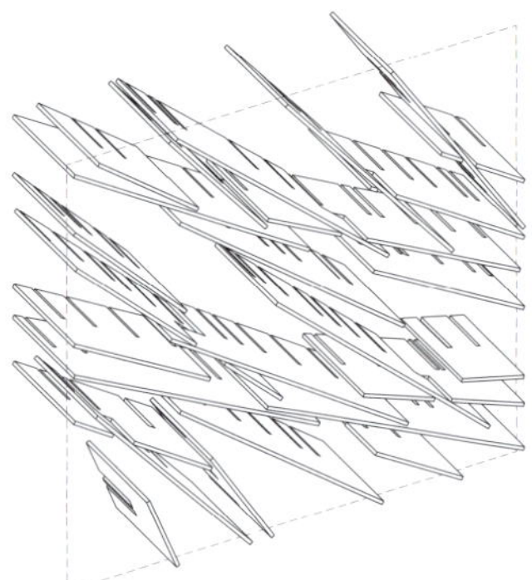

Direction A

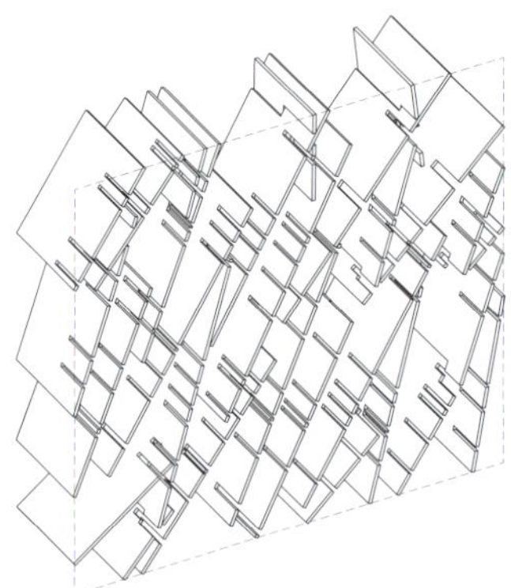

Direction B

Design Agency
Freeland Buck

Location
New York City, USA

Photographer
Kevin Kunstadt

SLIPSTREAM

Inspired by Lebbeus Woods' Slipstreaming drawings, the installation is made from over one thousand CNC cut plywood pieces that notch together to create an undulating, dynamically patterned and brightly colored wall. Developed as the extrusion of a 2-dimensional drawing through the gallery space, the structure is then cut away to produce a set of interconnected 3-dimensional spaces. The project develops novel forms of digital drawing, "egg-crate" type assemblies typical in stick built construction, and the designers' ability to describe and produce the dynamics of flow and turbulence, phenomena that have fascinated artists at least since Leonardo Da Vinci.

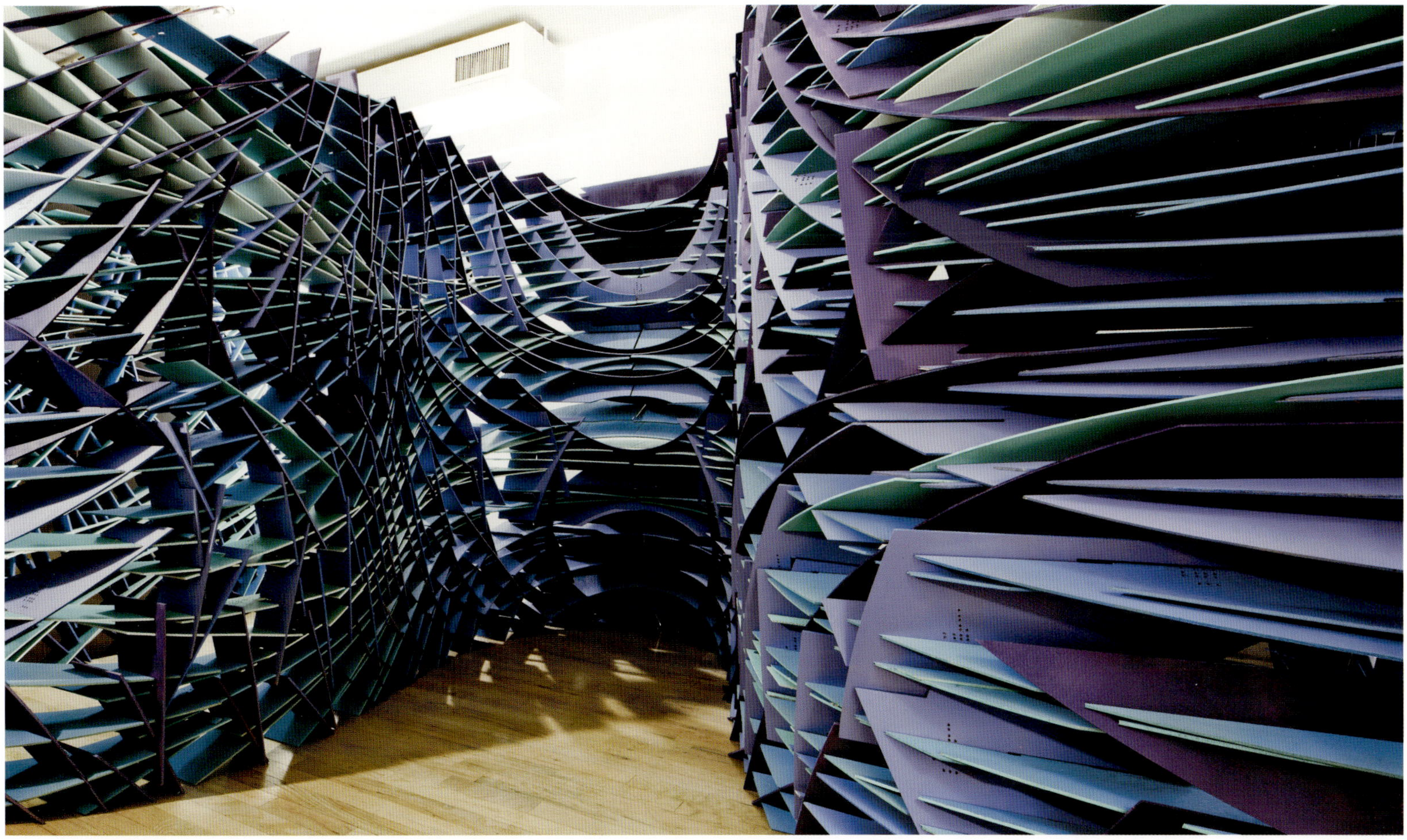

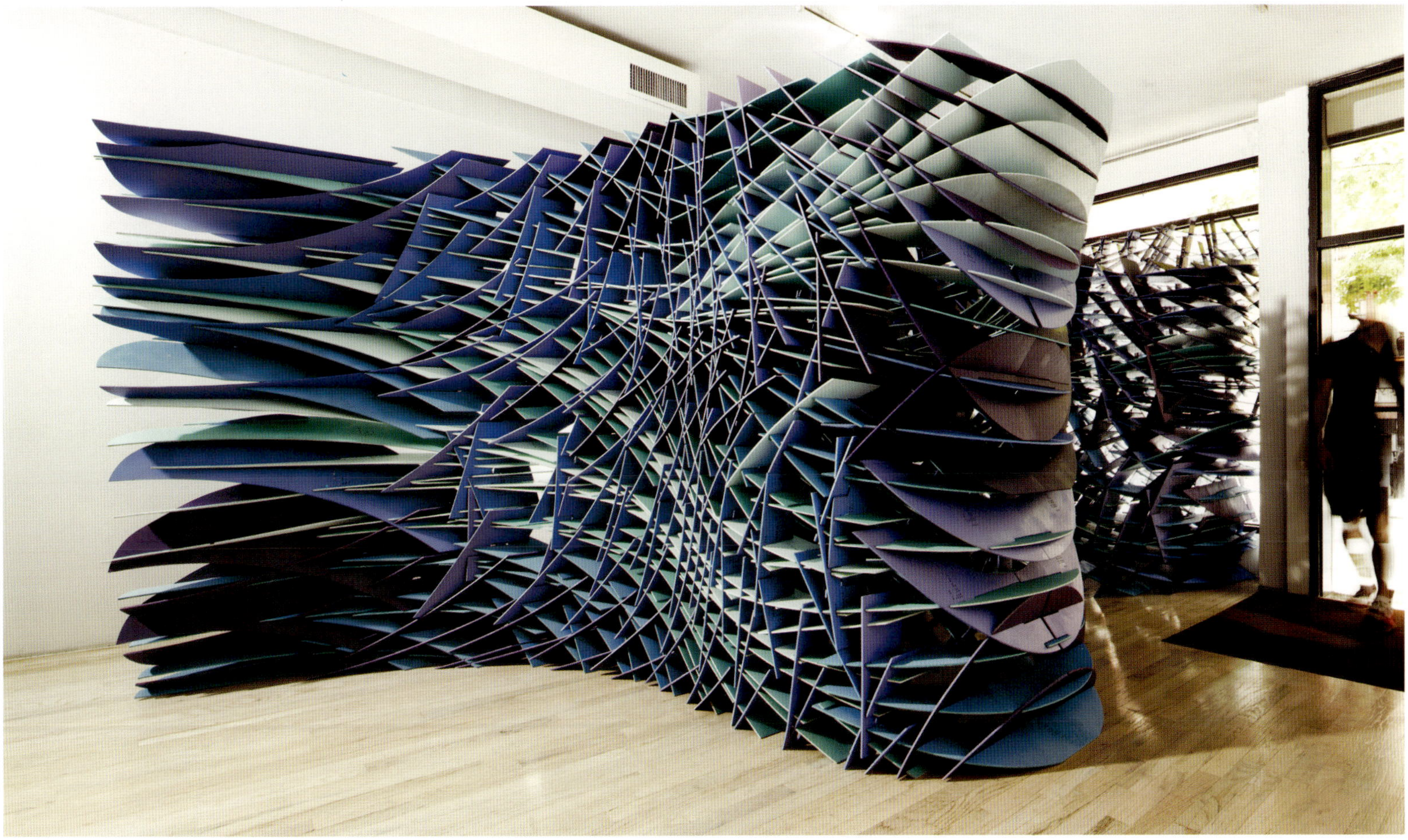

Lead Designer
Garth Britzman

Location
Lincoln, Nebraska, USA

Photographer
Chris Paulsen, Garth Britzman, Kelsey Loontjer, Matt Spohr, Peter Olshavsky

(POP)CULTURE

(POP)Culture was an installation that used recycled soda bottles to create canopy under which conversations about sustainability and urban land use can occur. The surface of the bottles creates an intriguing environment where one can explore the surface qualities of the bottles at eye level. The installation sought to draw attention to unique and unconventional uses for common goods. (POP)Culture used over 1,500 recycled plastic soda bottles and took 300 hours to create. It was displayed in Lincoln, Nebraska.

The lead designer of (POP)Culture was Garth Britzman and was fabricated with help from 26 students of the University of Nebraska Lincoln.

(POP)Culture was designed as an installation for Lincoln PARK(ing) Day in Lincoln, Nebraska, which was under the direction of Peter Olshavsky, Assistant Professor of Architecture at the University of Nebraska-Lincoln and made possible by support from the Fulbright Canada-RBC Eco-Leadership Program. Park(ing) Day is an annual international event. The temporary installation has been disassembled and recycled.

SERVICE AND DELIVERY VEHICLES ONLY

LOTUS CENTRAL DISPLAY

Six historic Lotus Formula 1 cars were incorporated in a spectacular sculpture by artist Gerry Judah that was the centrepiece of Goodwood Festival of Speed 2012, Britain's largest car culture event.

The 28-metre-tall sculpture was the 16th created by Gerry Judah for the Festival of Speed, an annual event held in the grounds of Goodwood House in West Sussex, the family seat of the Earl of March.

Each year, Goodwood has featured a marque, a carmaker that inspires disciples because of its style, success on the track, or both. In 2012, the featured marquee was Lotus, the British car manufacturer that sponsored Gerry Judah's installation.

The sculpture was designed to capture the essence of Lotus from its beginnings to the present. A 3D infinity loop was designed, resembling the grandest, most ambitious Scalextric track ever imagined.

The track itself is a triangular section of 6 mm flat sheet metal with a "continuously variable curve developable" surface, which was painted white. These were engineered by Capita Symonds and fabricated by Littlehampton Welding and transported 22 miles to Goodwood in 11 sections by individual articulated lorries and a police escort.

Multiple cranes were used to erect the installation and place six significant Lotus cars onto its surface. The cars, which were loaned by Classic Team Lotus and the Lotus F1 Team, included a green and yellow Type 32B, the car in which Jim Clark won the 1965 Tasman Series in Australia and New Zealand, and a red-and-white Type 49, in which Graham Hill raced to the crown. The other cars were a JPS-liveried Type 72, in

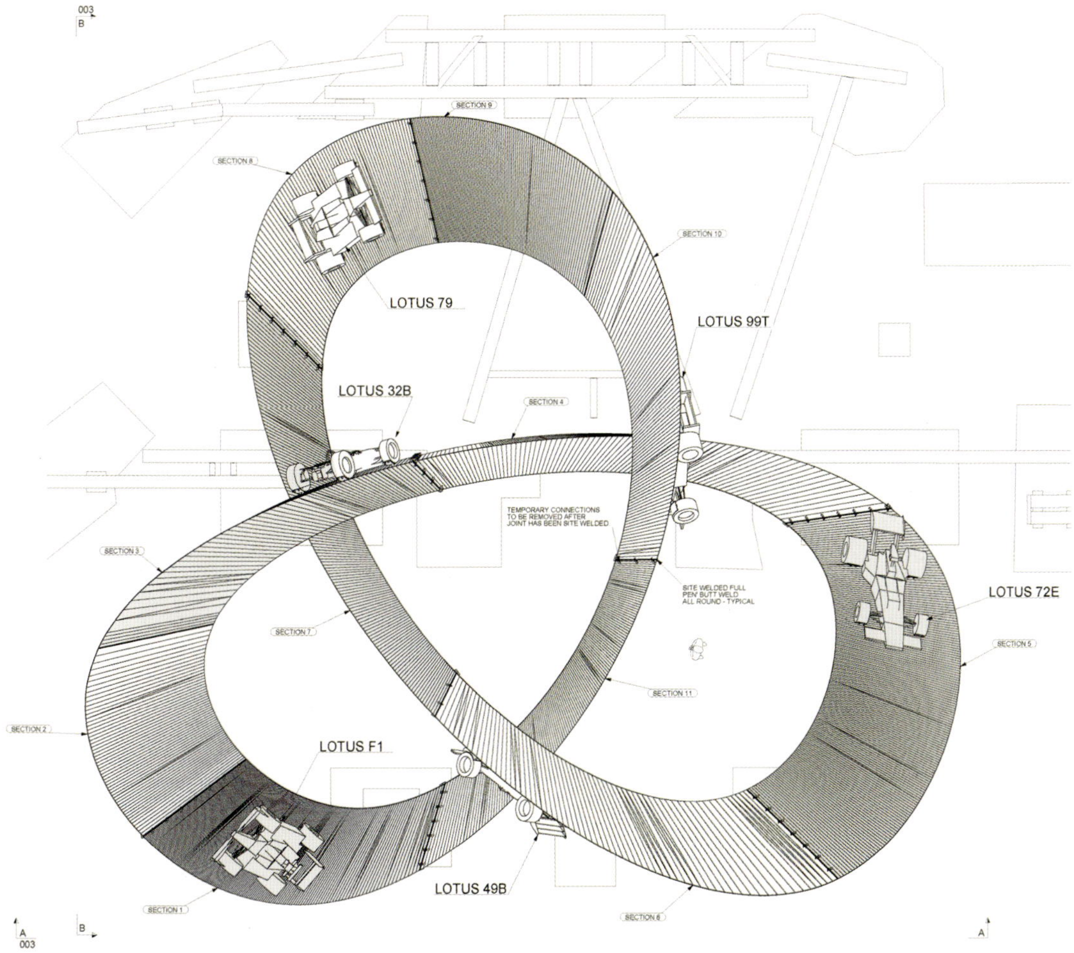

PLAN

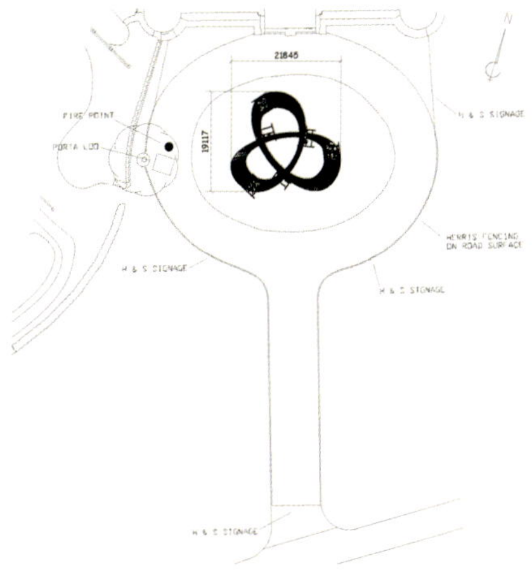

SITE LAYOUT PLAN

Designer
Gerry Judah

Client
Lotus Cars

Location
Goodwood Festival of Speed, UK

Engineering
Capita Symonds

Fabrication & Installation
Littlehampton Welding

Photographer
David Barbour

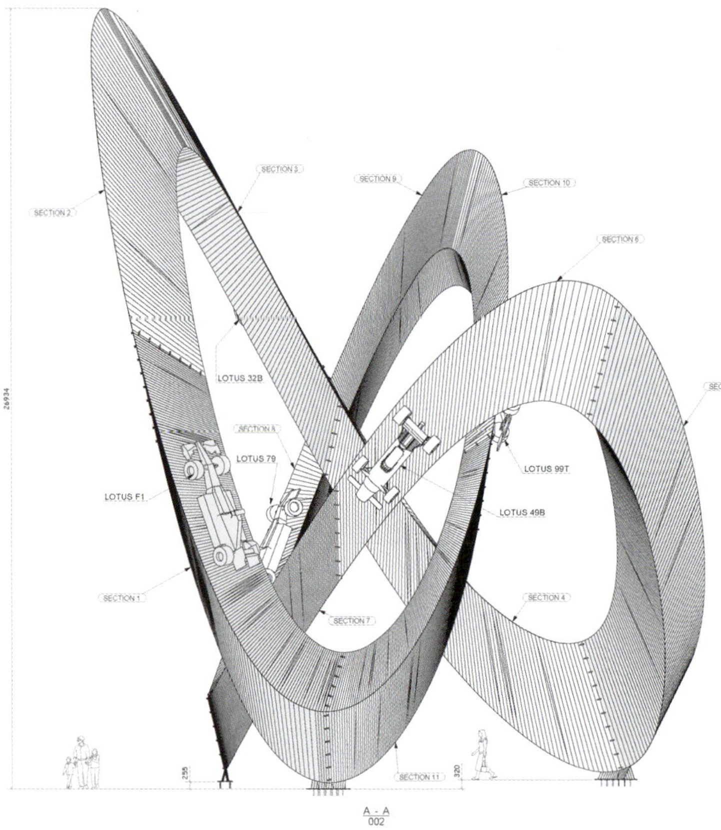
SECTION 2
SECTION 3
SECTION 9
SECTION 10
SECTION 6
LOTUS 32B
SECTION 5
SECTION 8
LOTUS 79
LOTUS F1
LOTUS 99T
LOTUS 49B
SECTION 1
SECTION 7
SECTION 4
SECTION 11
26934
255
320
A - A
002

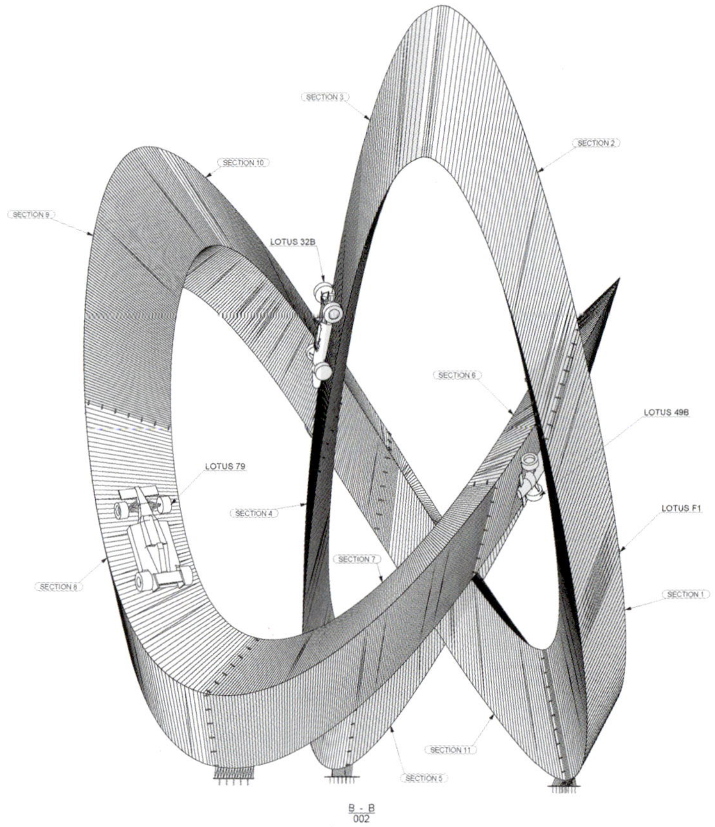
SECTION 3
SECTION 10
SECTION 2
SECTION 9
LOTUS 32B
SECTION 6
LOTUS 49B
LOTUS 79
SECTION 4
LOTUS F1
SECTION 7
SECTION 8
SECTION 1
SECTION 11
SECTION 5
B - B
002

which Emerson Fittipaldi became the sport's youngest champion; a black-and-gold '79 responsible for Mario Andretti's world title; a yellow Lotus 99T driven by Ayrton Senna; and the current Lotus grand prix car driven by Kimi Raikkonen and Romain Grosjean.

Gerry Judah worked closely with Lord March and Lotus to determine the design. The winding curves represent Lotus's natural environment: cars that are built for cornering. The 150-metre-long track is shaped into the shape of a half-hitch or trefoil-knot.

"What you see in the structure is the track, but inside it is 98% empty space," explains Gerry Judah. "In automobile terms, this would be a monocoque body, a tribute to the legendary designer and Lotus founder Colin Chapman's introduction of monocoque chassis construction to automobile racing."

"The monocoque structure, which is made of steel plates and joined together to create the loop, is meant to highlight the engineering DNA of Lotus," confirms Gerry Judah. "It's a lightweight engineering construction and I think its form shows the Lotus psychology and culture."

"What we have here is a technique for building freeform shapes. In the future, we expect that lots of structures will be built like this, from bridges and large span buildings, to roller coasters, but before that we will be building some even more spectacular sculptures."

ICD / ITKE RESEARCH PAVILLON 2010

Project Team
Institute for Computational Design – Prof. Achim Menges, Institute of Building Structures and Structural Design – Prof. Jan Knippers

Concept & Realisation
Andreas Eisenhardt, Manuel Vollrath, Kristine Wächter & Thomas Irowetz, Oliver David Krieg, Àdmir Mahmutovic, Peter Meschendörfer, Leopold Möhler, Michael Pelzer, Konrad Zerbe

Scientific Development
Moritz Fleischmann (project management), Simon Schleicher (project management), Christopher Robeller (detailing / construction management), Julian Lienhard (structural design), Diana D'Souza (structural design), Karola Dierichs (documentation)

Supporters
OCHS GmbH; KUKA Roboter GmbH; Leitz GmbH & Co. KG; A. WÖLM BAU GmbH; ES CAD Systemtechnik GmbH; Ministerium für Ländlichen Raum, Ernährung und Verbraucherschutz Landesbetrieb Forst Baden-Württemberg (ForstBW)

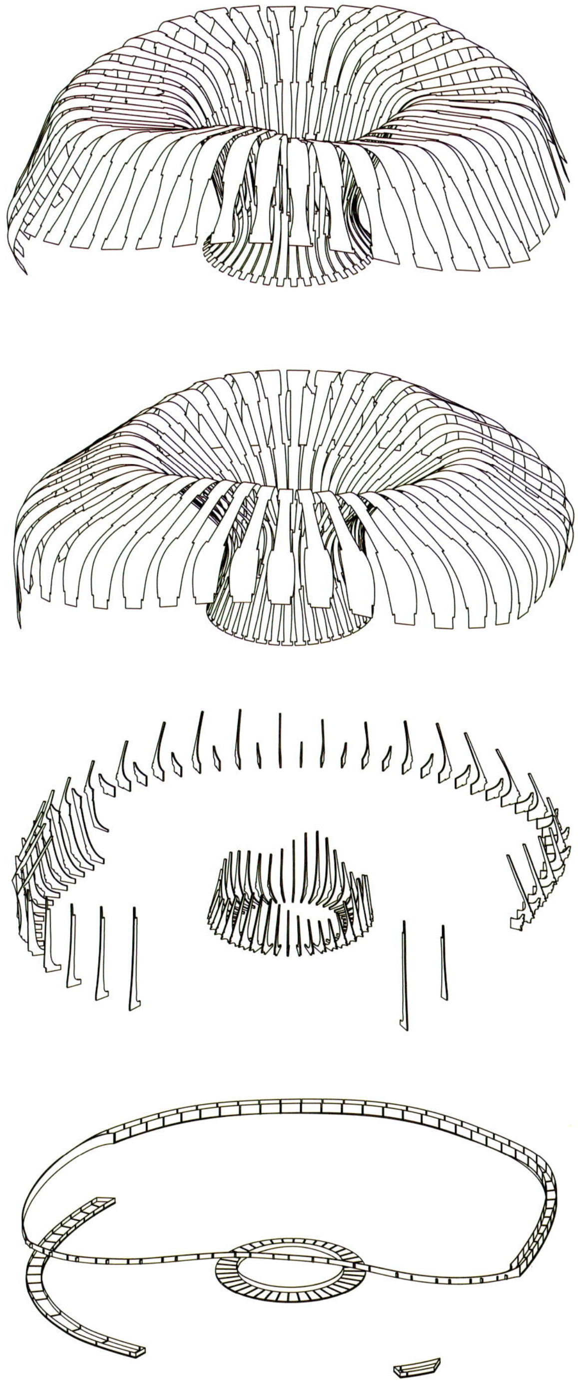

In July 2010, the Institute for Computational Design (ICD) and the Institute of Building Structures and Structural Design (ITKE), both at the University of Stuttgart, constructed a temporary research pavilion. The innovative structure demonstrates the latest developments in material-oriented computational design, simulation, and production processes in architecture. The result is a bending-active structure made entirely of elastically-bent plywood strips.

Any material construct can be considered as resulting from a system of internal and external pressures and constraints. Its physical form is determined by these pressures. However, in architecture, digital design processes are rarely able to reflect these intricate relations. Whereas in the physical world material form is always inseparably connected to external forces, in the virtual processes of computational design form and force are usually treated as separate entities, as they are divided into processes of geometric form generation and subsequent simulation based on specific material properties.

The research pavilion demonstrates an alternative approach to computational design: here, the computational generation of form is directly driven and informed by physical behavior and material characteristics. The structure is entirely based on the elastic bending behavior of birch plywood strips. The strips are robotically manufactured as planar elements, and subsequently connected so that elastically bent and tensioned regions alternate along their length. The force that is locally stored in each bent region of the strip, and maintained by the corresponding tensioned region of the neighboring strip,

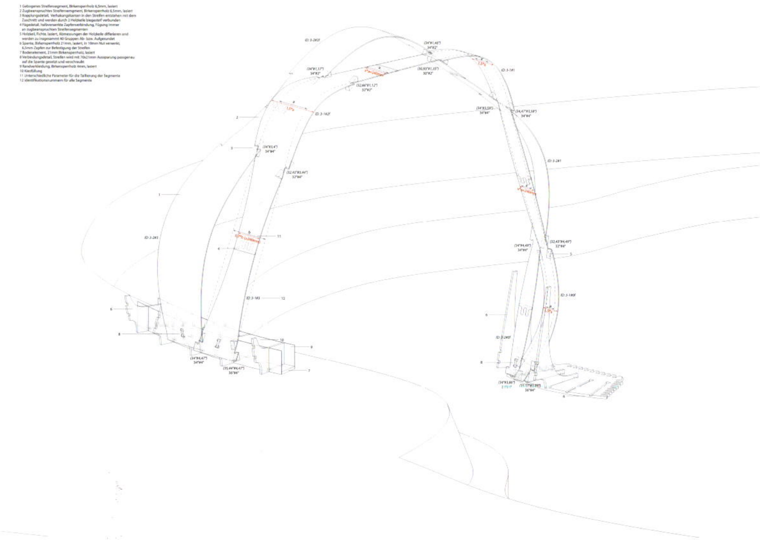

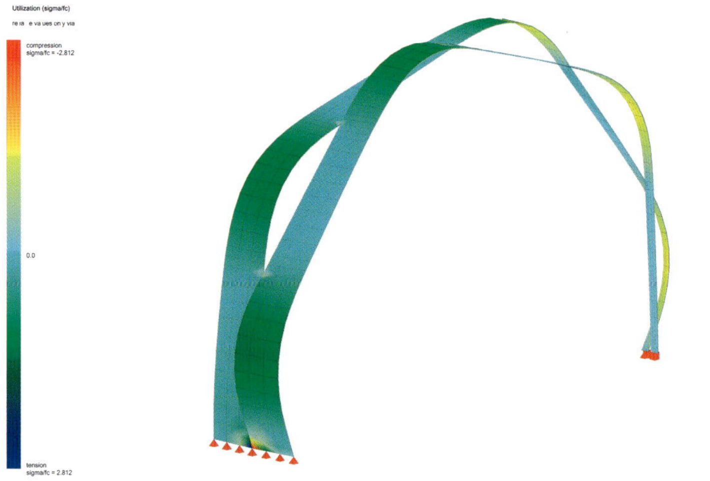
Utilization (sigma/fc)
compression
sigma/fc = -2.812
0.0
tension
sigma/fc = 2.812

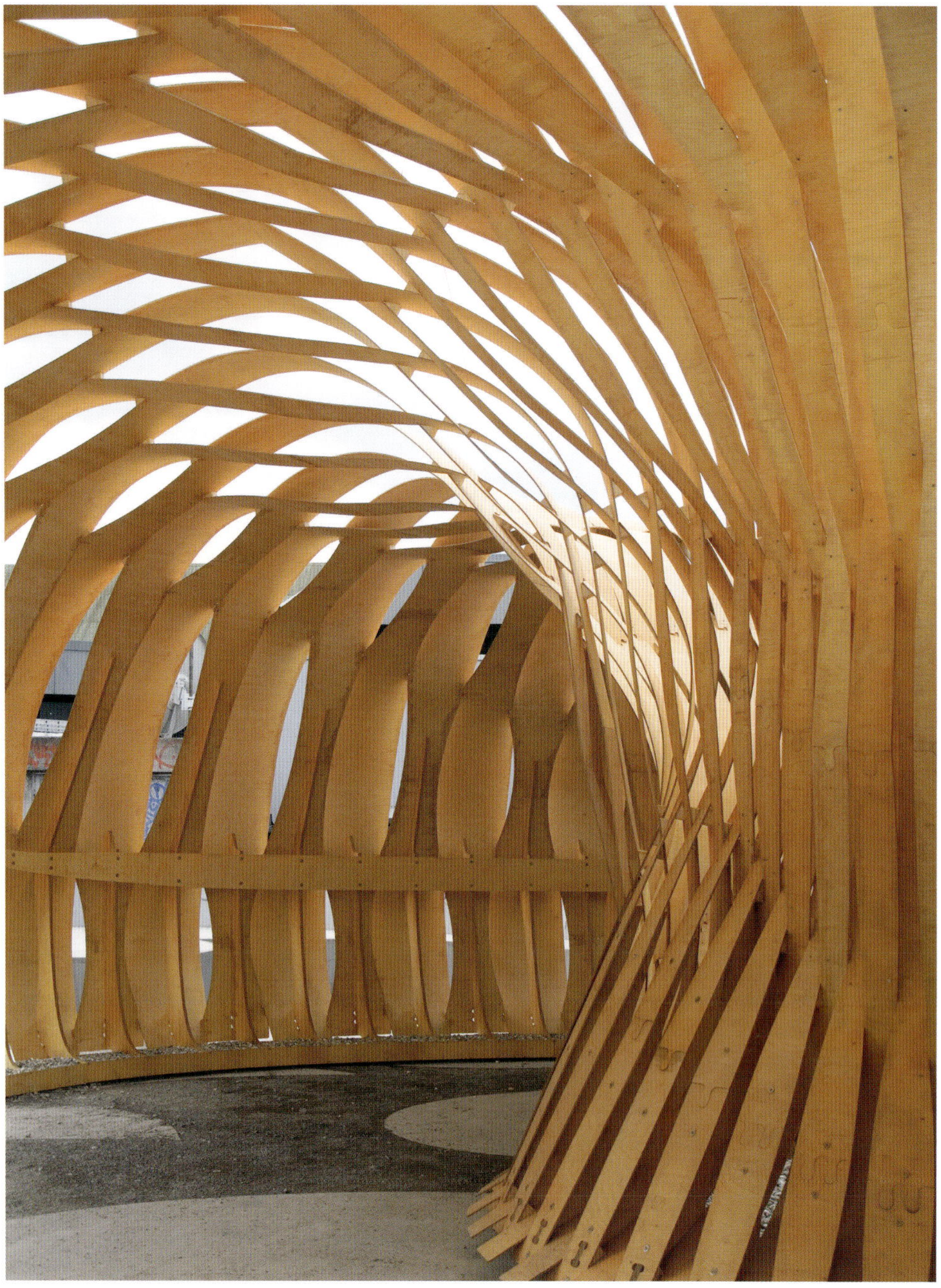

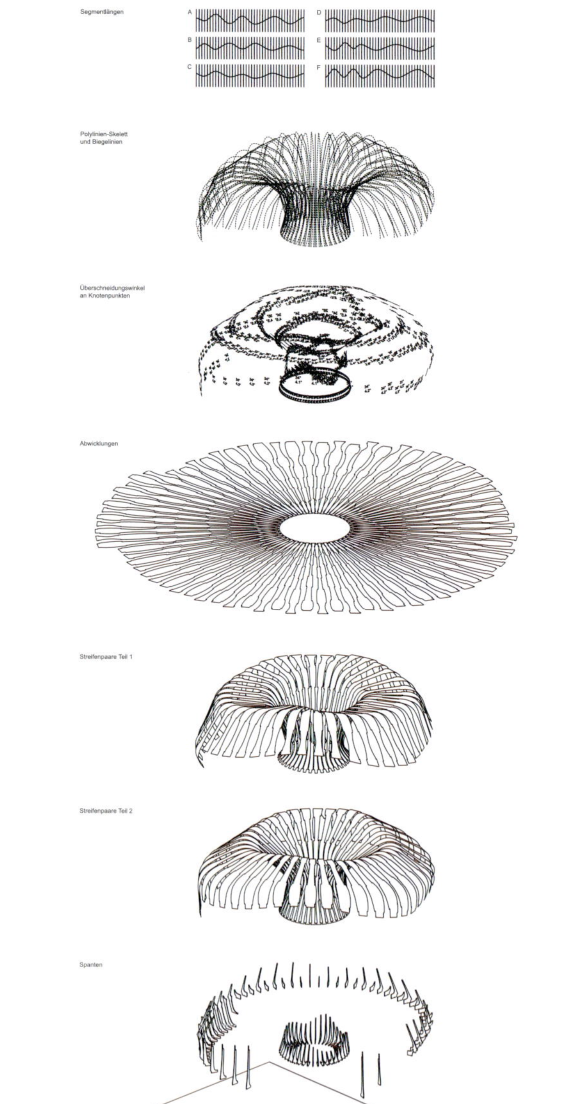
Segmentlängen
A
B
C
D
E
F
Polylinien-Skelett
und Biegelinien
Überschneidungswinkel
an Knotenpunkten
Abwicklungen
Streifenpaare Teil 1
Streifenpaare Teil 2
Spanten
Kieskasten
und Zugband

greatly increases the structural capacity of the system. In order to prevent local points of concentrated bending moments, the locations of the connection points between strips needs to change along the structure, resulting in 80 different strip patterns constructed from more than 500 geometrically unique parts. The combination of both the stored energy resulting from the elastic bending during the construction process and the morphological differentiation of the joint locations enables a very lightweight system. The entire structure, with a diameter of more than twelve meters, can be constructed using only 6.5 millimeter thin birch plywood sheets.

The computational design model is based on embedding the relevant material behavioral features in parametric principles. These parametric dependencies were defined through a large number of physical experiments focusing on the measurement of deflections of elastically bent thin plywood strips. Based on 6,400 lines of code one integral computational process derives all relevant geometric information and directly outputs the data required for both the structural analysis model and the manufacturing with a 6-axis industrial robot.

The structural analysis model is based on a FEM simulation. In order to simulate the intricate equilibrium of locally stored energy resulting from the bending of each element, the model needs to begin with the planar distribution of the 80 strips, followed by simulating the elastic bending and subsequent coupling of the strips. The detailed structural calculations, which are based on a specifically modeled mesh topology that reflects the unique characteristics of the built prototype, also allows for understanding the internal stresses that occur due to the bending of the material in relation to external forces such as wind and snow loads – a very distinct aspect of calculating lightweight structures.

Comparing the generative computational design process with the FEM simulation and the exact measurement of the geometry that the material "computed" on site demonstrates that the suggested integration of design computation into materialization is a feasible proposition.

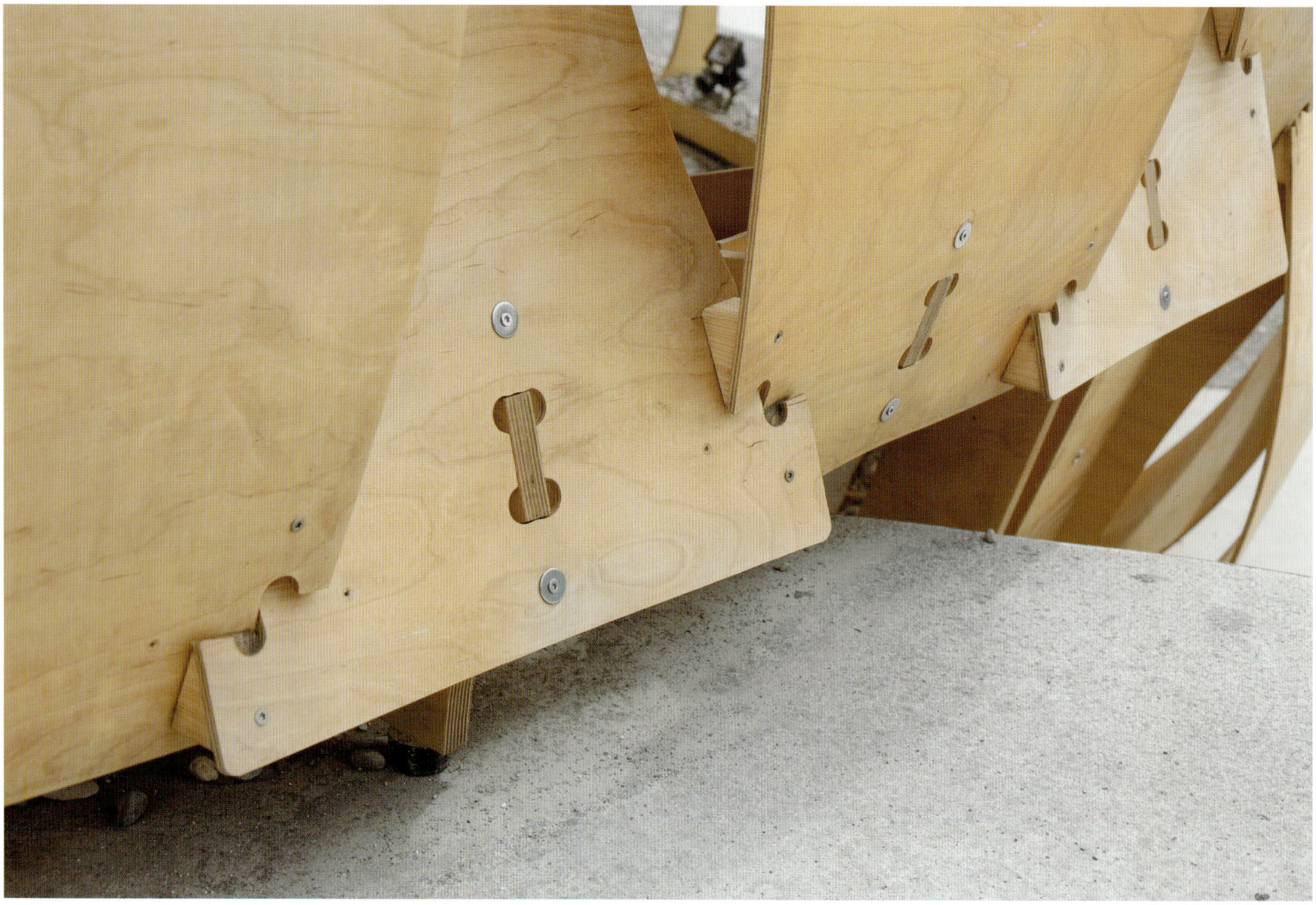

ICD / ITKE RESEARCH PAVILION 2011

In summer 2011 the Institute for Computational Design (ICD) and the Institute of Building Structures and Structural Design (ITKE), together with students at the University of Stuttgart have realized a temporary, bionic research pavilion made of wood at the intersection of teaching and research. The project explores the architectural transfer of biological principles of the sea urchin's plate skeleton morphology by means of novel computer-based design and simulation methods, along with computer-controlled manufacturing methods for its building implementation. A particular innovation consists in the possibility of effectively extending the recognized bionic principles and related performance to a range of different geometries through computational processes, which is demonstrated by the fact that the complex morphology of the pavilion could be built exclusively with extremely thin sheets of plywood (6.5 mm).

The project aims at integrating the performative capacity of biological structures into architectural design and at testing the resulting spatial and structural material-systems in full scale. The focus was set on the development of a modular system which allows a high degree of adaptability and performance due to the geometric differentiation of its plate components and robotically fabricated finger joints. During the analysis of different biological structures, the plate skeleton morphology of the sand dollar, a sub-species of the sea urchin (Echinoidea), became of particular interest and subsequently provided the basic principles of the bionic structure that was realized. The skeletal shell of the sand dollar is a modular system of polygonal plates, which are linked together at the edges by finger-like calcite protrusions. High load bearing capacity is achieved by the particular geometric arrangement of the plates and their joining system. Therefore, the sand dollar serves as a most fitting model for shells made of prefabricated elements. Similarly, the traditional finger-joints typically used in carpentry as connection elements, can be seen as the technical equivalent of the sand dollar's calcite protrusions.

Following the analysis of the sand dollar, the morphology of its plate structure was integrated in the design of a pavilion. Three plate edges always meet together at just one point, a principle which enables the transmission of normal and shear forces but no bending moments between the joints, thus resulting in a bending bearing but yet deformable structure. Unlike traditional lightweight construction, which can only be applied to load optimized shapes, this new design principle can be applied to a wide range of custom geometry. The high lightweight potential of this approach is evident

Architect
ICD / ITKE University of Stuttgart

Project Team
Institute for Computational Design – Prof. AA Dipl.(Hons) Achim Menges Achim Menges, Institute of Building Structures and Structural Design – Prof. Dr.-Ing. Jan Knippers, Competence Network Biomimetics Baden-Württemberg

Planning and Realisation
Peter Brachat, Benjamin Busch, Solmaz Fahimian, Christin Gegenheimer, Nicola Haberbosch, Elias Kästle, Oliver David Krieg, Yong Sung Kwon, Boyan Mihaylov, Hongmei Zhai

Concept and Project Development
Oliver David Krieg, Boyan Mihaylov

Scientific Development
Markus Gabler (project management), Riccardo La Magna (structural design), Steffen Reichert (detailing), Tobias Schwinn (project management), Frédéric Waimer (structural design)

Location
Stuttgart, Germany

Surface
72 m^2

Volume
200 m^3

Material
275 m^2 Birch plywood 6.5 mm Sheet thickness

Photographer
ICD / ITKE University of Stuttgart

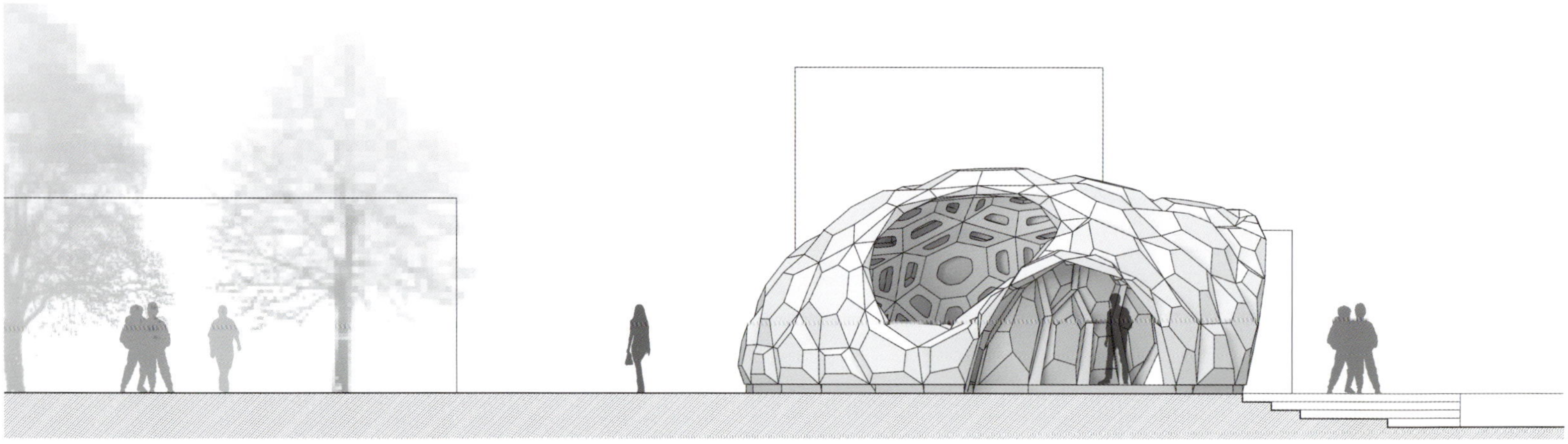

as the pavilion that could be built out of 6.5 mm thin sheets of plywood only, despite its considerable size. Therefore it even needed anchoring to the ground to resist wind suction loads.

Besides these constructional and organizational principles, other fundamental properties of biological structures are applied in the computational design process of the project, including the Heterogeneity, Anisotropy, and Hierarchy.

A requirement for the design, development and realization of the complex morphology of the pavilion is a closed, digital information loop between the project's model, finite element simulations and computer numeric machine control. Form finding and structural design are closely interlinked. An optimized data exchange scheme makes it possible to repeatedly read the complex geometry into a finite element program to analyze and modify the critical points of the model. In parallel, the glued and bolted joints were tested experimentally and the results included in the structural calculations.

The plates and finger joints of each cell were produced with the university's robotic fabrication system. Employing custom programmed routines the computational model provided the basis for the automatic generation of the machine code (NC-Code) for the control of an industrial seven-axis robot.

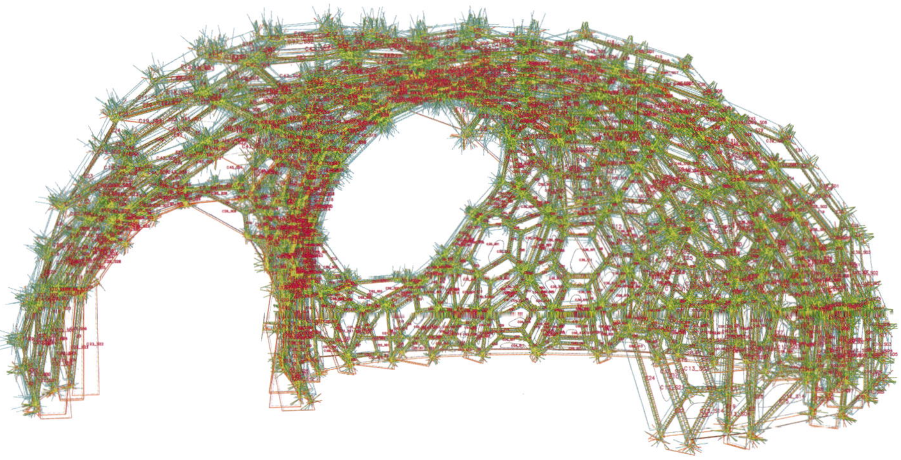

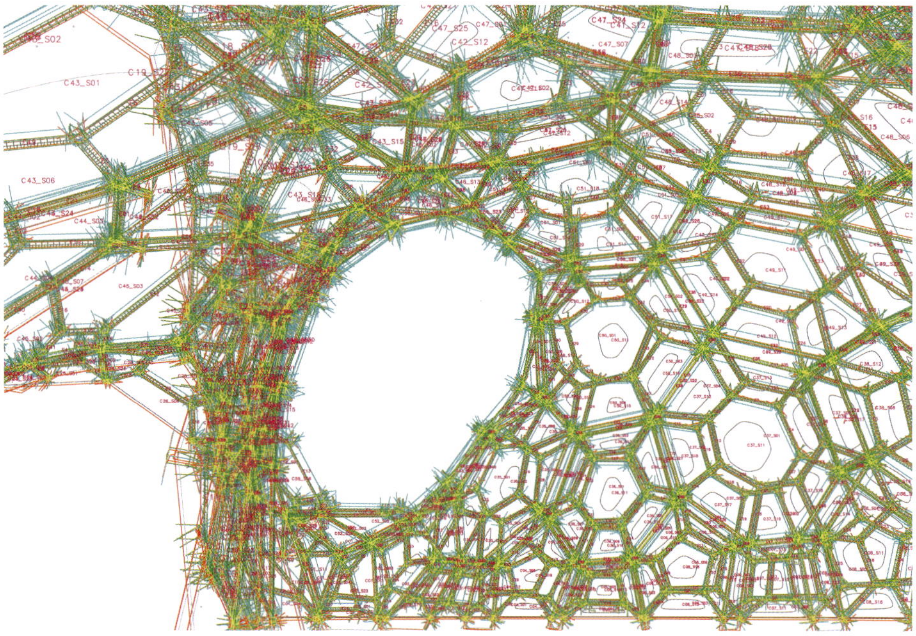

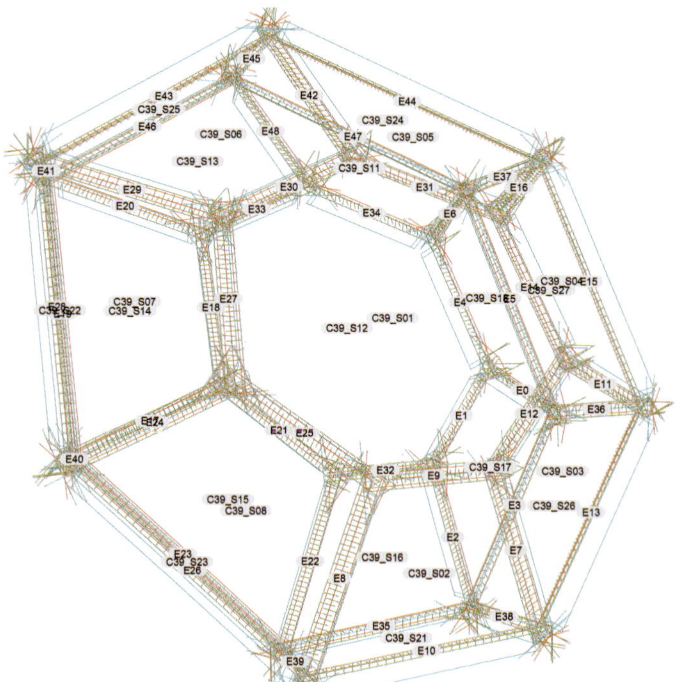
E45
E43
C39_S25
E46
E42
E44
C39_S24
C39_S06
E48
E47
C39_S05
C39_S13
C39_S11
E41
E29
E20
E30
E33
E31
E37
E16
E34
E6
C39_S07
C39_S14
E18
E27
C39_S01
C39_S12
E4
E5
C39_S04
E15
C39_S27
E0
E11
E1
E12
E36
E21
E25
E40
E32
E9
C39_S17
C39_S03
C39_S15
C39_S08
E3
C39_S26
E13
E2
E7
E22
E8
C39_S16
C39_S02
E23
C39_S23
E26
E35
E38
C39_S21
E10
E39

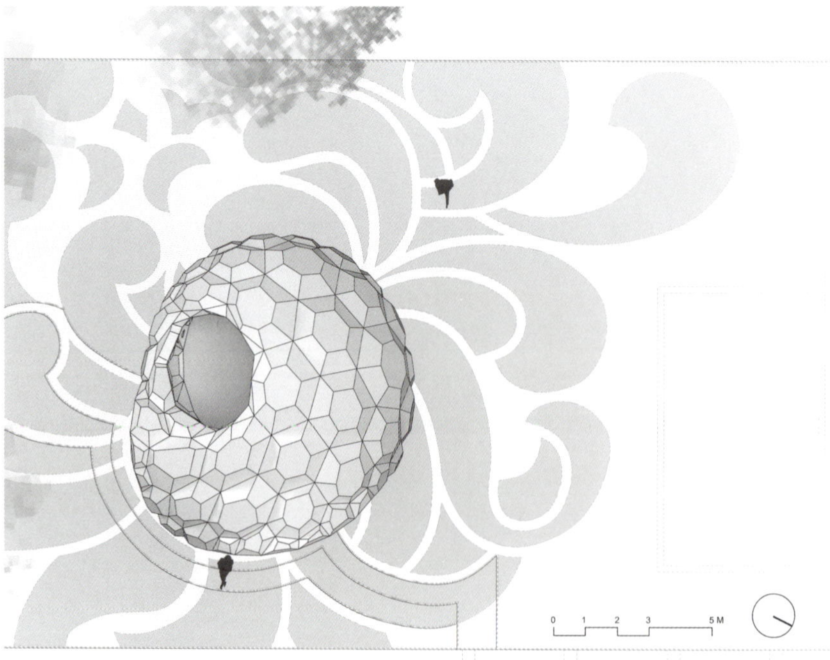
0 1 2 3 5 M

DRAGON SKIN PAVILION

The Dragon Skin Pavilion is an architectural installation designed and built for the 2011-12 Hong Kong & Shenzhen Bi-City Biennale of Urbanism\Architecture.

The Pavilion utilizes a newly developed environmentally friendly material called "post-formable" plywood, which incorporates layers of adhesive film to allow easy single-curved bending without the need for steam or extreme heat. With no material loss, a CNC mill divided 21 of these 8x4 plywood sheets into eight identical squares, and accurately cut the unique connection slots that were programmed into the pavilion geometry by computer. Using one single mould, all panels were bent into the same shape, and within six hours the numbered shells were slotted into place without using any plan drawings, glue or screws. The underlying equilibrium surface geometry removed all internal forces and deformations from the pavilion, which became a self-supporting, free-standing, light-weight skin with highly tactile tectonic properties and unique lighting effects.

The structure challenges and explores the spatial, tactile, and material possibilities that architecture can offer by revolutions in digital fabrication and manufacturing technology. The Dragon Skin Pavilion redefines the role of architectural design in construction by actively working with the material's basic properties and pushing its structural performance, while being aware of the aesthetic values and effects the system provides.

The pavilion is the product of a collaboration between the Laboratory for Explorative Architecture & Design (LEAD) and EDGE Laboratory for Architectural and Urban Research (Tampere University of Technology, Finland). It was designed by KristofCrolla (LEAD), Sebastien Delagrange (LEAD), Emmi Keskisarja (EDGE), and PekkaTynkkynen (EDGE) and builds upon expertise from a first prototype constructed during an architectural design workshop "Material Design & Digital Fabrication" at the Tampere University of Technology.

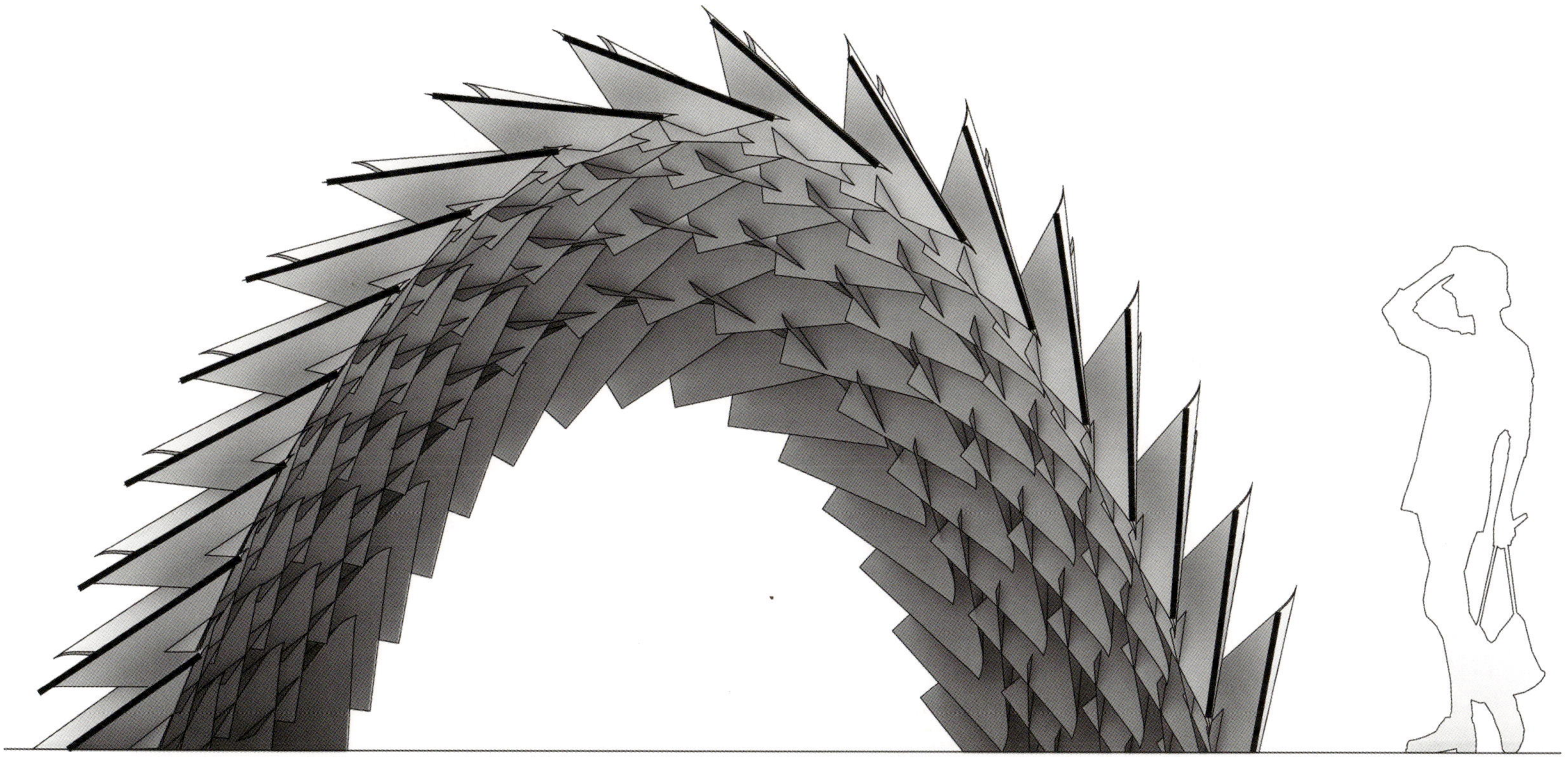

Design Agency

EDGE Laboratory for Architectural and Urban Research, Tampere University of Technology, School of Architecture, Laboratory for Explorative Architecture & Design Ltd. (LEAD)

Design Team

KristofCrolla (LEAD), Sebastien Delagrange (LEAD), Emmi Keskisarja (EDGE), PekkaTynkkynen (EDGE)

Client

2011-12 Hong Kong & Shenzhen Bi-city Biennale of Urbanism\ Architecture

Location

Kowloon Park, Hong Kong, China

Photographer

Dennis Lo Designs

GOLDEN MOON

The Golden Moon is a temporary architectural structure that explores how Hong Kong's unique building traditions and craftsmanship can be combined with contemporary design techniques in the creation of a highly expressive and captivating public event space. It is the 2012 Gold Award winning entry for the Lantern Wonderland design competition organised by the Hong Kong Tourism Board for the Mid-Autumn Festival and was on display for 6 days in Hong Kong's Victoria Park.

Traditional materials for making lanterns, such as translucent fabric, metal wire and bamboo, have been translated to a large scale. A light-weight steel geodesic dome forms the pavilion's primary structure and is the basis for a computer-generated grid wrapped around it. This grid is materialised through a secondary structure from bamboo. For this, Hong Kong's traditional bamboo scaffolding techniques were used – a high-speed, instinctive way of building scaffoldings for e.g. the city's many skyscrapers. This highly intuitive and imprecise craft was merged with exact digital design technology to accurately install and bend the bamboo sticks into a grid wrapping the steel dome. This grid was then clad with stretch fabric flames, all lit up by animated LED lights.

The bamboo and flames follow a pattern based on an algorithm for sphere panellisation that produces purity and repetition around the equator and imperfection and approximation at the poles. This gradual change, combined with the swooping and energetic curves that define the geometry, creates a very dynamic space that draws spectator's view up towards the tip. By putting the axis of this cladding grid not vertical but under an angle, the dome gets an asymmetric directionality. This motion is reinforced by the entrance which is placed along this tilted axis to draw people into the sphere and where they get swept away along the grid's tangents and vectors. The colouration of the pavilion amplifies this effect of submergence in a light wonderland. On top of the black painted steel structure, which forms a neutral base, eight different, saturated colours of stretch fabric are used for the flames. The colours gradually range from ivory and yellow to intense orange, red and deep bordeaux. The brightest colours are used at the tilted base whereas the darkest colours are used at the pole where they, together with the more scrambled geometry, make the pattern disintegrate into the black night sky.

The Golden Moon builds up on research into "building simplexity", the building of

Competition Design
Kristof Crolla of LEAD and Adam Fingrut

Project Management
Laboratory for Explorative Architecture & Design Ltd. (LEAD)

Location
Hong Kong, China

Photographer
Kevin Ng, Grandy Lui and Pano Kalogeropoulos

complex geometry and space using the simplest of means. In this research we strategically combine digital design techniques, such as computer programming or CNC fabrication, with traditional crafts and basic materials. In this project procedural modelling techniques were used to control the production of the unique geometry: a sphere that is wrapped with a diagrid according to a Fibonacci sequence that produces order along the equator and randomness at the poles. Code was used for the production of simple drawings that would allow the labour force to mark up intersections between the steel structure and bamboo easily. These drawings took traditional bamboo scaffolding construction detailing into consideration in the definition of installation tolerances. Optimisation scripts were finally used to reduce the amount of unique stretch-fabric "flames" from 470 different units to 10 different types that could stretch and adapt to the various conditions in which they were applied. All details and construction procedures were devised to allow for a high-speed production as only 11 days of onsite construction were available for this 6-storey-high pavilion. To bring the project to a successful end within the limited time available, a very close conversation with the craftsmen was required from the beginning. Preconceptions of building methods and familiar construction techniques had to be abandoned by all parties as both the digital and the material world demanded a new design and building set-up to be devised. This project shows an alternative way for digital design to be materialised into a more humane environment with real-world conditions like limited time frames, low budgets, minimal precision but human flexibility, creativity and ad-hoc inventiveness.

LEDSCAPE

"LEDscape" is an installation which deals with light as a constructive element of space and landscape. It is located in the "Centro Cultural de Belém" in Lisbon and aims to introduce the everyday user to the LEDARE light bulb, thus demystifying the preconceived ideas about LED technology and emphasizing the relevance of this product in the sustainability of the future. "LEDscape" challenges the passerby to an interactive experience — a pathway gradually lit with 1,200 light spots, invites introspection and individual appropriation of the installation.

Architect
LIKEarchitects

Location
Belém Cultural Center, Lisbon, Portugal

Photographer
FG+SG – Fernando Guerra, Sergio Guerra

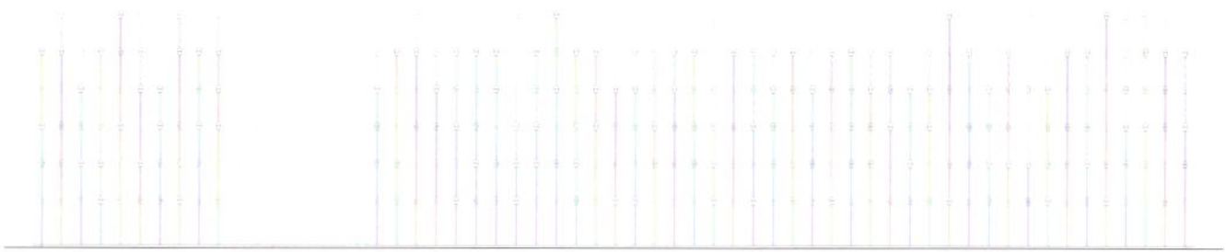

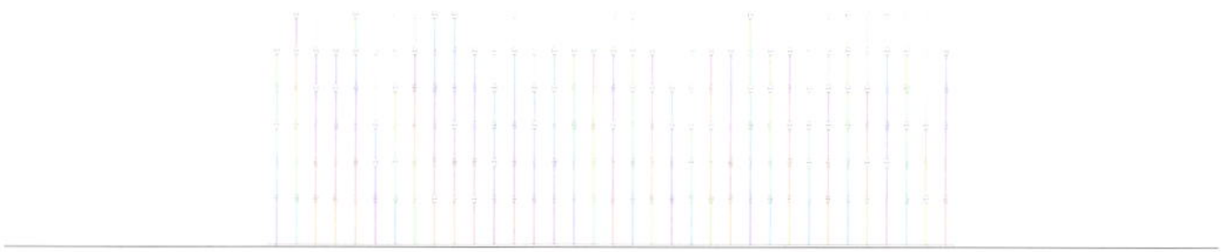

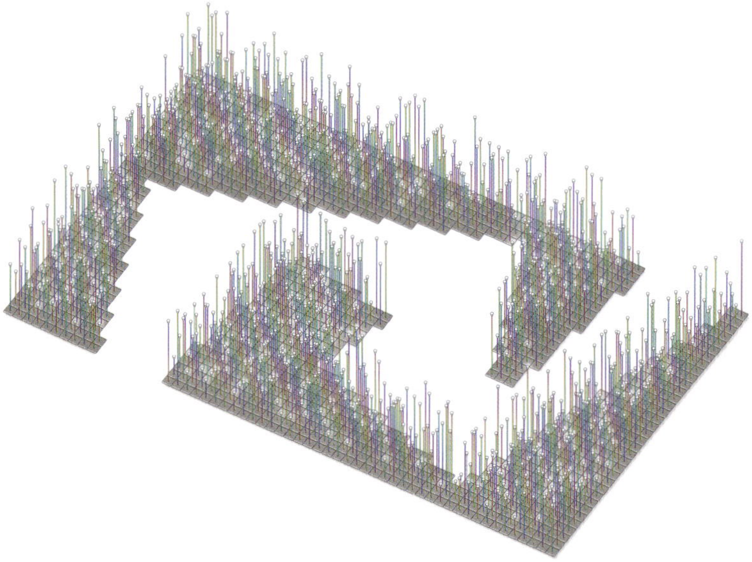

CHROMATIC SCREEN

Chromatic Screen states LIKEarchitects presence in the space of "Point of View" in Oporto Show'12, by invitation of "design factory".

For this annual design show, the Portuguese architects' collective designed an intervention representative of their (ephemeral) work that lies on the border between Architecture, Design, Urban Installation and Art.

Starting from the (re) interpretation of a common object of daily use, Chromatic Screen builds a distinct element, spatial organizer, that involves all who visit it into a chromatic experience with different moments of opacity and transparency. It is designed from 2000 hangers for children's clothes from IKEA "Bagis", in 4 different colours — blue, green, pink and orange – that merge into multiple tonalities, Chromatic Screen results in a kinetic and very colourful experience.

Designer
LIKEarchitects / Diogo Aguiar + Teresa Otto + João Jesus

Location
Edifício da Alfândega, Porto, Portugal

Principal Use
Art Instalation

Principal Materials
2000 hangers for children's clothes – 'Bagis' (polypropylene plastic)

Curatorship
design factory*

Dimensions
1.5 m x 1.5 m x 2 m

Installation Height
2 m

Production
Francisco Andrade, Metalomecânica

Construction
LIKEarchitects

Photographer
Dinis Sottomayor

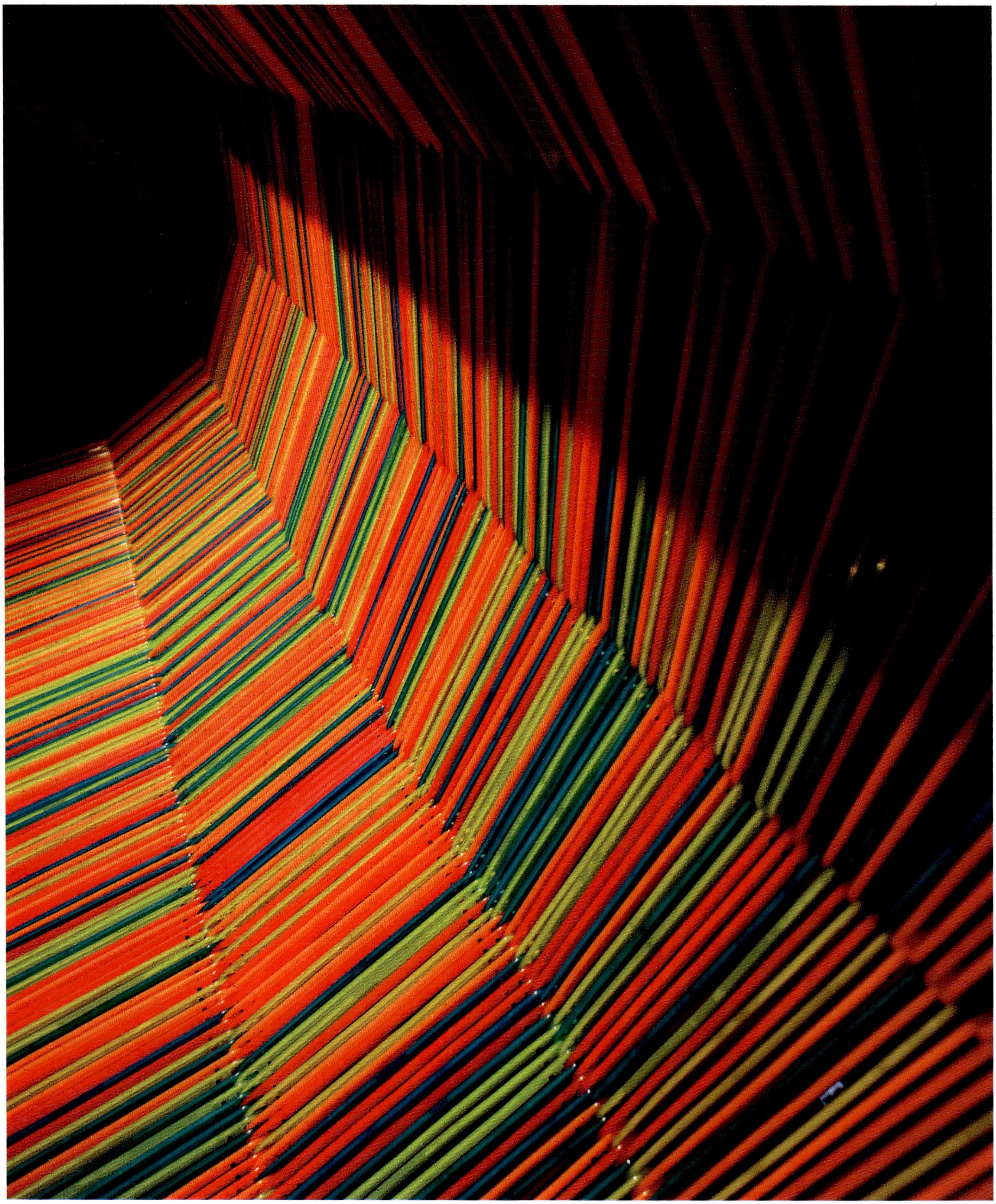

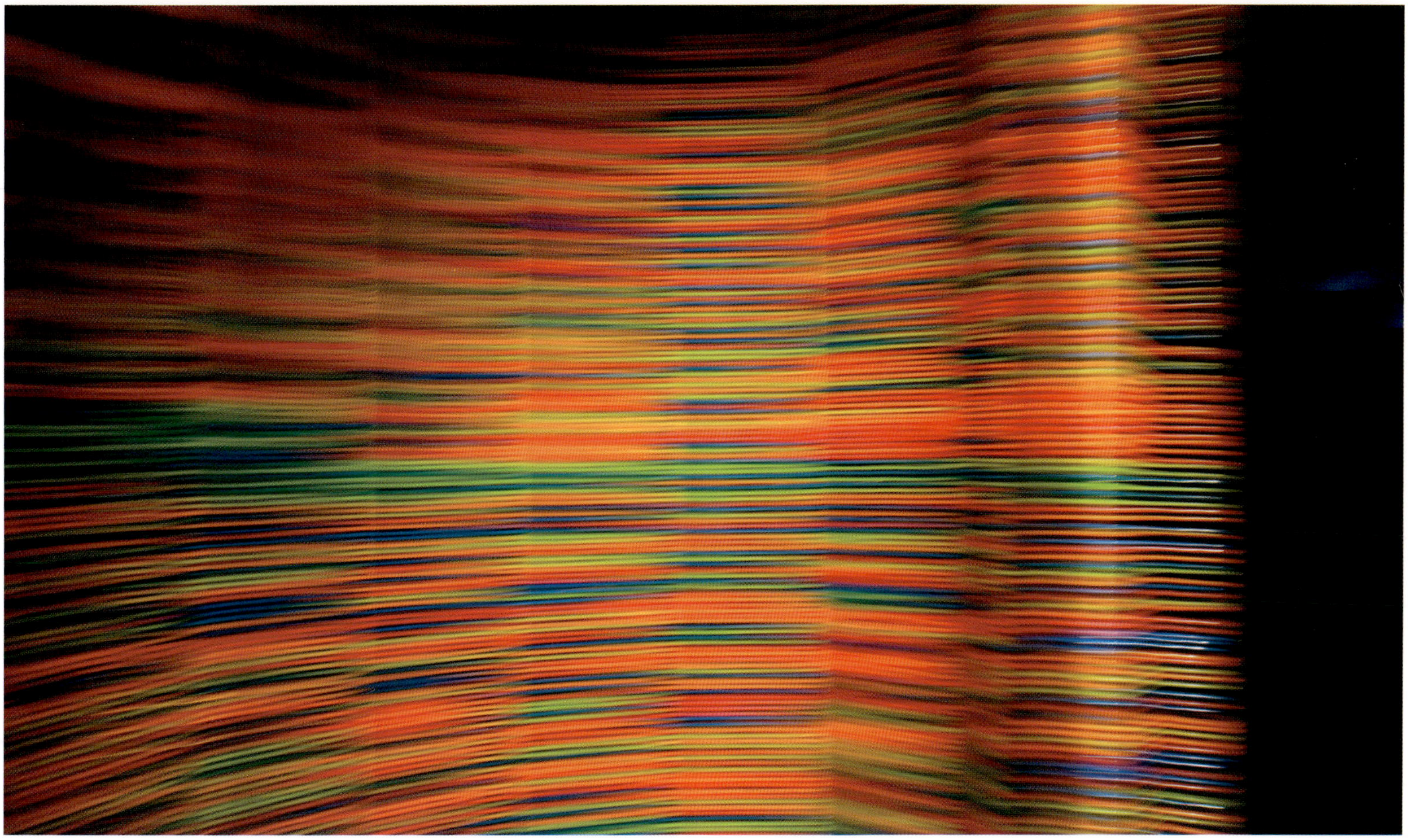

Design Agency
Los Carpinteros

Artist
Marco Castillo, Dagoberto Rodriguez

Photographer
Roberto Chamorro

GÜIRO: AN ART BAR INSTALLATION

Güiro, an art bar installation is conceived as Gesamtkunstwerk. Los Carpinteros's construction draws on the concept of an open-air art bar inspired by the Güiro, a ubiquitous Cuban percussion instrument made from a dried hard-shell, tropical fruit. It is also the word used in Cuban slang to connote a party.

Oval in shape, the slatted structure was lit from within to emphasize its gridded design. Its rectangular openings functionedas seating for visitors, while the centerwas occupied by a round bar complete with cocktails designed by the artists.

A collaboration between Absolut Art Bureau and Los Carpinteros for Art Basel Miami Beach, the installation was open on 5 – 9th December 2012, and was accompanied by a curated program of live music.

In addition to its utilitarian function and musical references, Güiro expanded on a body of work by Los Carpinteros which fashions civic spaces from the architectural language of panopticon prisons. The distinct circular configuration of the panopticon prison was developed in the 18th century by the English philosopher Jeremy Bentham and allowed a centrally placed guard to watch numerous prisoners placed along the perimeter. Another work in this series was Sala de Lectura Ovalada (2011) — a ten-foot-high reading room — on view at Sean Kelly Gallery in New York in the spring of 2011.

OOSTCAMPUS

In 2008 the Flemish Government Architecture Agency (Vlaams Bouwmeester) announced an international competition for ideas to build OostCampus, with a slogan that paraphrases Magritte: "Ceci n'est pas … een Administratief centrum".

The winning project, by the Madrid studio led by Carlos Arroyo, opted for a radical re-use of the large industrial existing building, including foundations, floors, supporting structures, outer skin, insulation, waterproofing, and all recoverable services and equipment: power station, heating plant, water pipes, fire hoses, sewerage, and even parking area, fencing and access.

The reuse of the existing is a basic criterion of sustainability. The "gray energy" (energy used for the production of something) is often discarded or simply ignored. If we demolish an existing structure and build again, we will use more energy and resources than the most efficient of buildings can save in its life span.

To transform the vast industrial hall – with minimal footprint but maximum spatial result – Arroyo designs a sheltered interior public space, wrapped in a "luminous landscape of white clouds". Thin shells of GRG (gypsum and fibre) span the large space like huge soap bubbles. They are only 7 mm thick.

Within this landscape, a set of modular clusters provides the administrative services and spaces, designed to facilitate the relationship between citizens and administration. Citizen participation in the process is one of the key issues. Also transparency: the chamber hall is in full sight in the middle of public space, the information is accessible, you can even visualize the municipal website … and physically enter it and talk to the person who is behind!

POORT 7
POORT 6
POORT 5
POORT 4
POORT 3
POORT 2
POORT 1
0
5
10
20m

burgerzaken
Attesten
Adres
Burgerlijke stand
Identiteit
Reispassen
Rijbewijzen
Strafregister
Wonen
Werken
Welzijn
Burgerzaken
vrije tijd
OPENINGSUREN
CONTACTINFO
oostkamp.be

OCMW
Welzijn
3

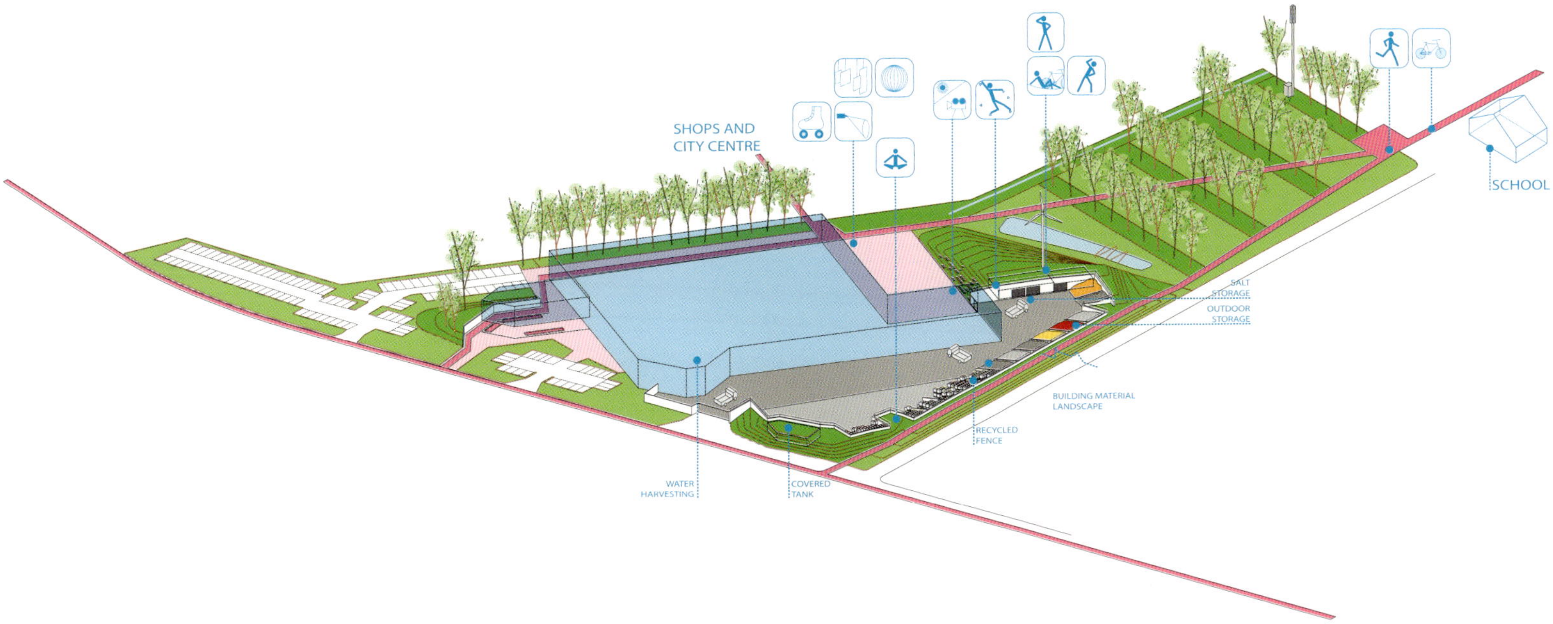

The materials are simple and inexpensive, but are selected and used in such a way that the designers want to go and touch them. Some elements are finished with a felt made from recycled bottles (PET); simple boards are CNC carved to become sophisticated 3D damascene; the floor is the existing industrial warehouse poli-concrete, with its lines of storage, on which the new signage is superimposed. The acoustics are carefully worked out, and so is the smell!

Thermal comfort is achieved with minimum effort, thanks to the technique of the "thermal onion" which optimizes climate areas according to levels of access, and making use of the thermal inertia of the concrete slab.

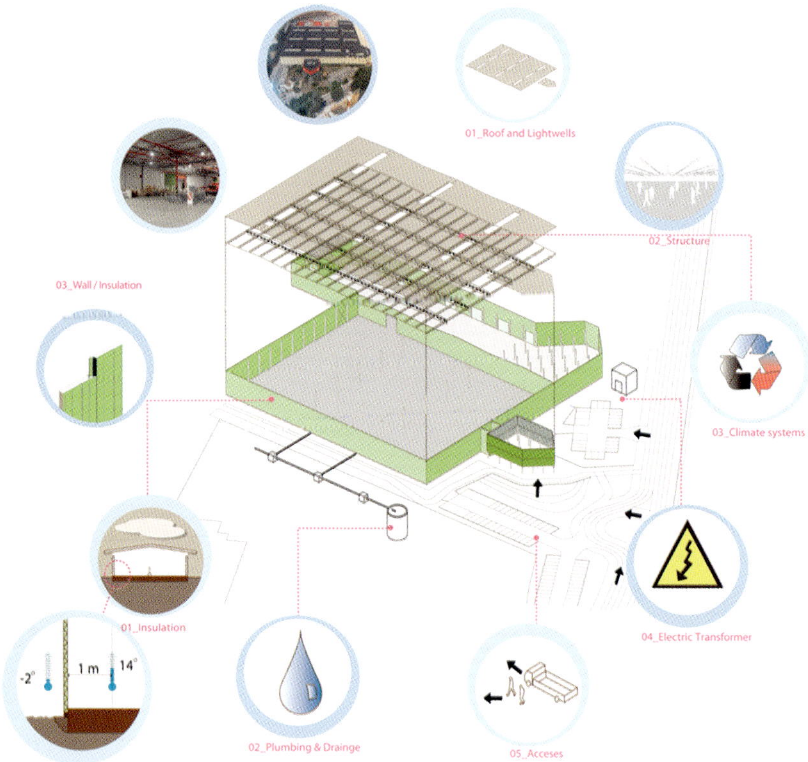

Architect
Carlos Arroyo

Project Leaders
Carlos Arroyo and Vanessa Cerezo

Design Team
David Berkvens, Carmina Casajuana, Irene Castrillo, Miguel Paredes, Benjamin Verhees, Pieter Van Den Berge, Luis Salinas, Sara Miguelez, Sarah Schouppe

Project Development
Wolkenbouwer (Carlos Arroyo Arquitectos, Spain + ELD Partnership, Belgium)

Client
Autonoom Gemeetebedrijf Oostkamp

Location
Oostkamp, Belgium

Area
11,000 m^2

Photographer
Miguel de Guzmán

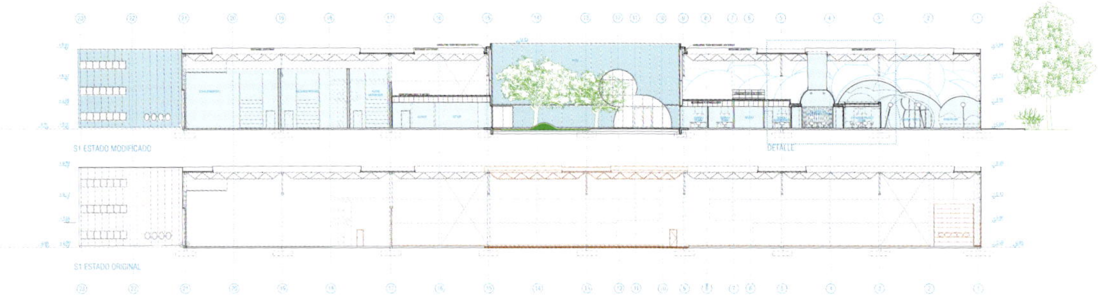
S1 ESTADO MODIFICADO
DETALLE
S1 ESTADO ORIGINAL

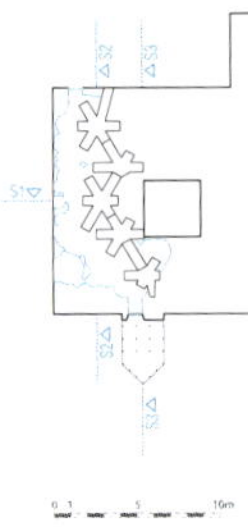

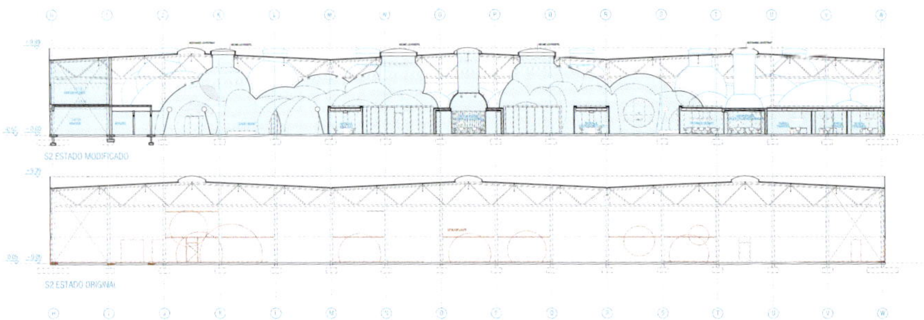
S2 ESTADO MODIFICADO
S2 ESTADO ORIGINAL

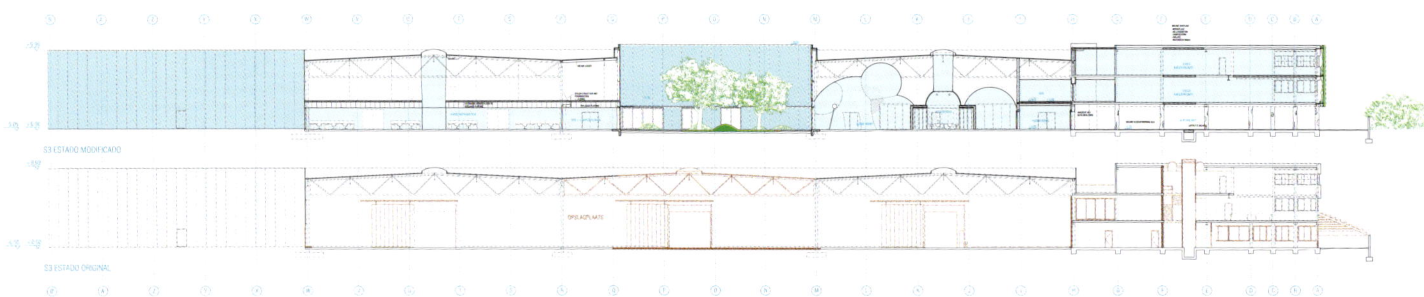
S3 ESTADO MODIFICADO
S3 ESTADO ORIGINAL

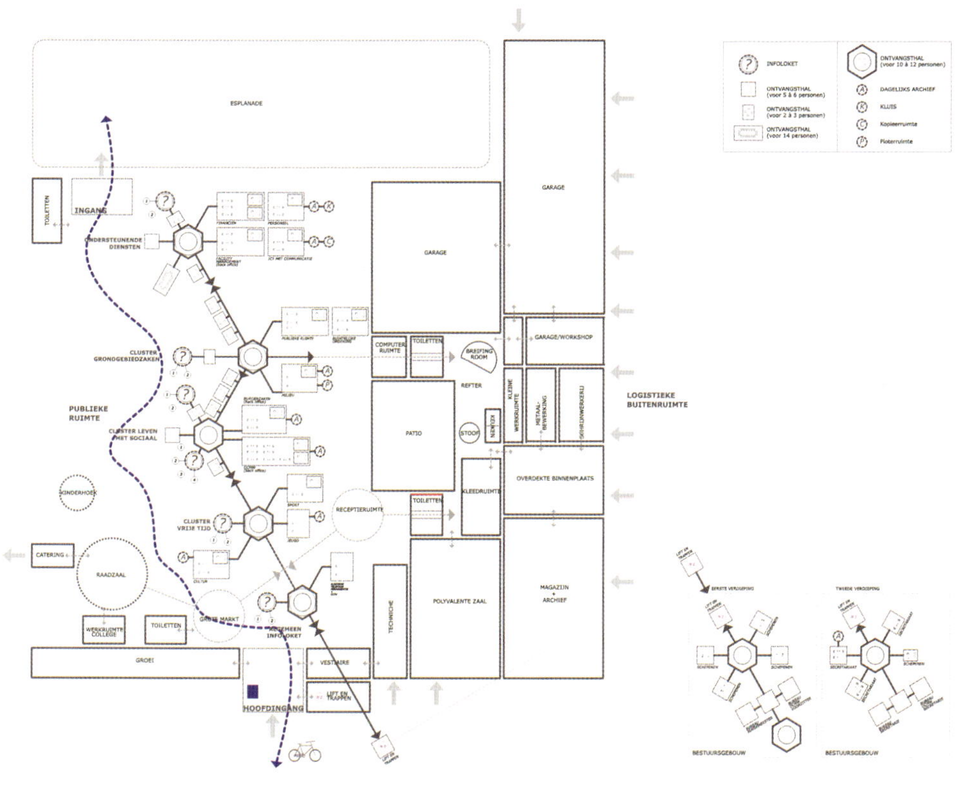
ESPLANADE
GARAGE
TOILETTEN
INGANG
ONDERSTEUNENDE DIENSTEN
CLUSTER GRONDGEBIEDZAKEN
PUBLIEKE RUIMTE
CLUSTER LEVEN MET SOCIAAL
KINDERHOEK
CLUSTER VRIJE TIJD
CATERING
RAADZAAL
GROTE MARKT
ALGEMEEN INFOLOKET
WERKRUIMTE COLLEGE
TOILETTEN
GROEI
HOOFDINGANG
VESTIAIRE
LIFT EN TRAPPEN
TECHNIEK
POLYVALENTE ZAAL
MAGAZIJN + ARCHIEF
RECEPTIERUIMTE
TOILETTEN
KLEEDRUIMTE
PATIO
STOOF
KEUKEN
REFTER
COMPUTER RUIMTE
TOILETTEN
BRIEFING ROOM
GARAGE/WORKSHOP
KLEINE WERKRUIMTE
METAAL-BEWERKING
SCHRIJNWERKERIJ
OVERDEKTE BINNENPLAATS
LOGISTIEKE BUITENRUIMTE
INFOLOKET
ONTVANGSTHAL (voor 5 à 6 personen)
ONTVANGSTHAL (voor 2 à 3 personen)
ONTVANGSTHAL (voor 14 personen)
ONTVANGSTHAL (voor 10 à 12 personen)
DAGELIJKS ARCHIEF
KLUIS
Kopieerruimte
Ploterruimte
BESTUURSGEBOUW
BESTUURSGEBOUW

CLIENIA KLINIK LITTENHEID - SPATIAL AND COMMUNICATION CONCEPT

A castle offers protection, every kid knows that! This is especially true for "Lino Castle," because in this case it concerns the psychiatric child and adolescent clinic at the Clienia clinic village in Littenheid in Eastern Switzerland. Here children and adolescents with depression, fear of school, ADHS, borderline disorders and traumas are treated – here they find help, safety and protection. The castle gates were open for the first time for younger and older patients in Augusr, 2012. Together with the children and adolescents from the clinic as well as doctors, care-providers, trained experts and clinic directors, dan pearlman developed with "Lino Castle" a spatial and communications concept that strongly promotes the recovery process of the children and adolescents and lets them feel at ease in their temporary home for the duration of their stay. From the strategic planning and the visual look, to the implementation of the interior design and furnishings concept, dan pearlman played the role of strategic partner to the clinic during all phases. With "Lino Castle" Clienia sets a standard for a lasting design of clinics not only in Switzerland but far beyond its borders.

For a period of months strategists and designers developed for Cliena the complete story of "Lino Castle" which is deeply rooted in the history of the clinic village and also reflects all the standard medical histories of the young patients. The five characters developed especially for Clienia – Alfa, Dipsta, Betha, Elysee and Calibri – function as guardian protectors of the individual wards and represent various types of ailments and emotional states. Each figure is distinguished by an individual colour palette as well as special traits. Binoculars, a steering wheel and a treasure chest are for example the faithful companions of Betha the Pirate. Instead of going to school, he would prefer to conquer the high seas and signalizes empathy, safety and understanding to the children and adolescents on his ward. Even Lino, the castle dragon and for whom the castle is named, has left his marks all over the clinic in the form of paw prints and scratch marks.

The architectural spatial concept also grew out of the storyline of Lino Castle and unites colours, forms, materials, graphics and furniture to create a healing and anxiety-free environment. Playful, inviting and something different than clinic grey, lovingly designed relaxation areas, interactive room elements and common areas await the young patients. The newly designed spatial and communications concept strongly supports the effectiveness of treatment, making starting therapy easier and increasing its success for the children and adolescents involved as well as their parents. As a result, Clienia conveys with "Lino Castle" a feeling of safety and understanding. Not least, the concept also allows Clienia to better position itself in the Swiss health care market. Never-before realized in Switzerland, this concept has received a wealth of attention from experts and the media.

WILLKOMMEN
Ich bin Lino, der Burgdrache. Ich bin sehr scheu. Auch wenn ihr mich nicht seht, ich bin immer für euch da. Wenn sich abends goldene Sonnenuntergänge am Himmel zeigen, wisst ihr, dass ich vor Freude Feuerwolken gespuckt habe. Und wenn morgens die Wiesen feucht sind, wisst ihr, dass auch ein Drache mal traurig sein kann. In dieser Burg habe ich neue Freunde gefunden. Sie helfen, die täglichen Abenteuer zu überstehen. Lasst es euch gut gehen mit euren Schutzpatronen Alfa, Betha, Calibri, Dipsta und Ely.
AHOI

Design Agency
dan pearlman Markenarchitektur GmbH

Designer
Volker Katschinski (Spatial Concept), Jork Andre Dieter (Communication Concept)

Architect
Karin Hechinger

Client
Clienia Klinik Littenheid

Location
Littenheid, Switzerland

Area
1,915 m^2

Photographer
Guido Leifhelm

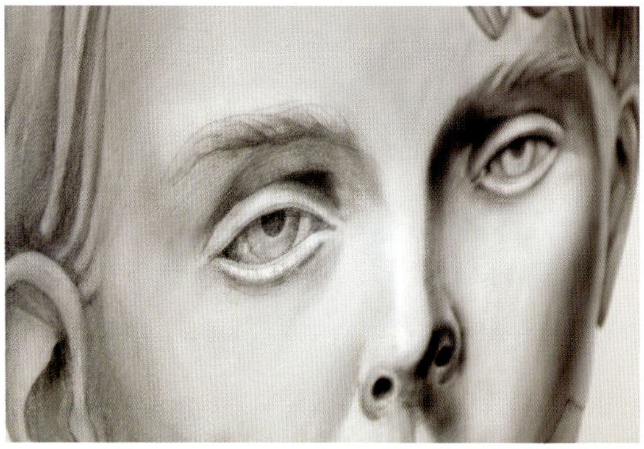

Design Agency
Dopludó Collective

Designer
Egor Kraft

Client
Artek Club

Location
Saint Petersburg, Russia

ARTEK MURAL

Hand drawn mural made in a bar, club and exhibition space 'Artek', in St. Petersburg, is an additional part of the interior, made by Lesha Galkin, the designer's teammate at Dopludo, as well as an independent artwork itself. The name 'Artek' is taken from a famous in post-Soviet states summer camp, located on the Crimean Peninsula. So the topic of the mural comes from that fact. Three stages of destruction of the bust of the pioneer, a typical soviet boy scout, is a metaphor for the collapse of the ambitions of the camp, that was supposed to be an international place of the national feuds, political differences and spiritual emptiness. Ambitions were dashed with the ambitions of the former Soviet Union. In fact in the third stage of the mural in the center of the head is a round ball, like a core, and also there're fragments of the crashed head that cast shadows on it, so that they form shapes of the continents of the earth on the core. This image symbolizes the idea of cosmopolitanism as a mean against political pressure, nationalism and involvement in the far-fetched processes.

BIT HOTEL

Hand drawn mural realized in the art influenced Hotel "Bit", located in Barcelona.

In carrying out the realization of the mural participated all teammates of Dopludó Collective: Karina Eybatova, Lesha Galkin & Egor Kraft.

Design Agency
Dopludó Collective

Client
Beriestain Interiores

Location
El Poblenou, Barcelona, Spain

Photographer
Dopludó Collective

I TRAVELED
ROUND
THE WORLD
LOOKING
FOR A
HOME

BETTER PLACE / COPENHAGEN

Designer Agency
Francisco Sarria / Studio

Designer
Francisco Sarria

Area
1,000 m^2

The permanent exhibition of over 1,000 m^2 conveys the core values of Better Place: namely that innovation and sustainability can go hand in hand and that zero-emission vehicle powered by electricity from renewable sources is the best way forward.

The main concept of the visitor centre is the use of green technology shaped by green interior architecture. The exhibition is a spectacular and engaging experience that aims to entertain and surprise rather than just inform its audience. The space was designed to create an emotive trip that invites visitors to explore, play and learn.

Using camera tracking and directional sound, interactive mirror cubes change from mirror mode to screen mode displaying animated content about the use of electric vehicles.

Core objectives of Better Place and information related to electric vehicles is displayed on a wooden wall by a number of RGB LED lights that illuminate the sentences creating a sequence of messages.

A cylindrical cinema for up to 20 people is one of the main features of the space. The cinema is operated by a wire system which hoist the cylindrical wall up and down. The cinema and all the audio visual elements of the exhibition are operated remotely.

Interactive quizzes to test the guests' knowledge after touring the visitor centre, touch screens and models illustrating the components of the Better Place infrastructure are all part of the exhibition.

Apart from conveying the Better Place messages, the installations and interior architecture have a functional purpose, providing furniture and spaces for visitors to meet, play around or just relax.

better place

better place

better place

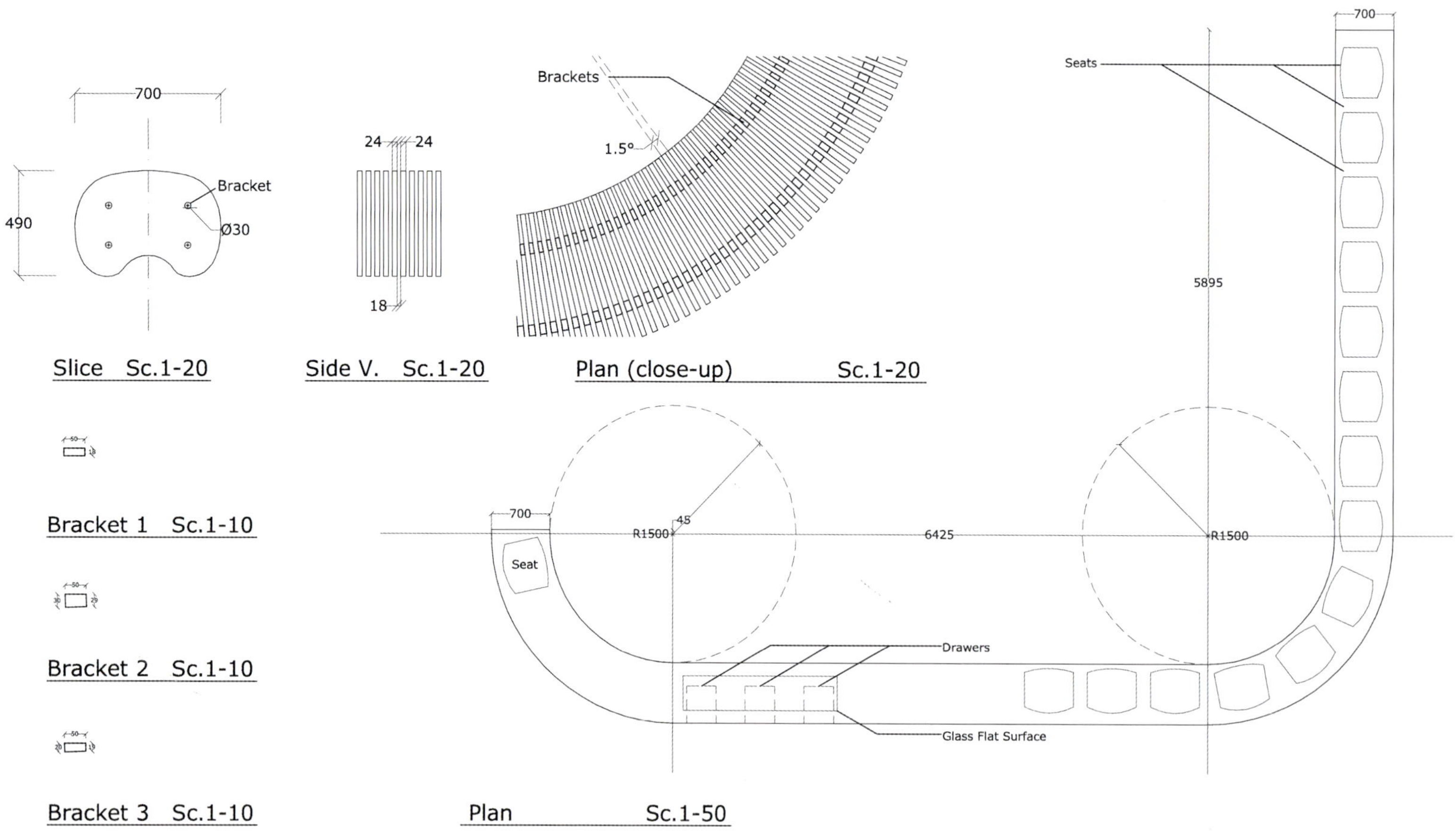
700
490
Bracket
Ø30
Slice Sc.1-20
24
24
18
Side V. Sc.1-20
Brackets
1.5°
Plan (close-up) Sc.1-20
Seats
700
5895
Bracket 1 Sc.1-10
Bracket 2 Sc.1-10
Bracket 3 Sc.1-10
700
Seat
R1500
45
6425
R1500
Drawers
Glass Flat Surface
Plan Sc.1-50

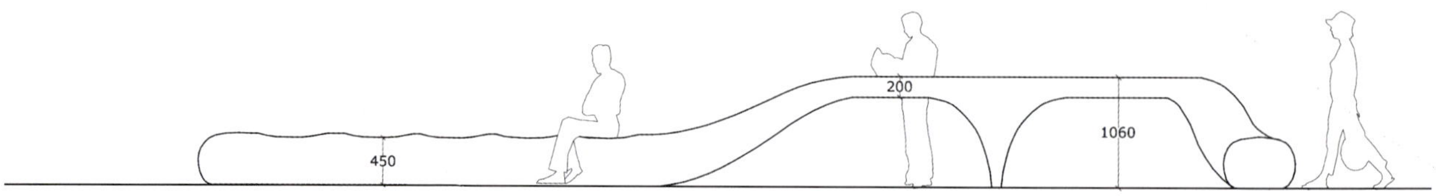
200
1060
450

are you ready to

better place

Zero
emission
Zero
worries
go electric now!
Zero
noise

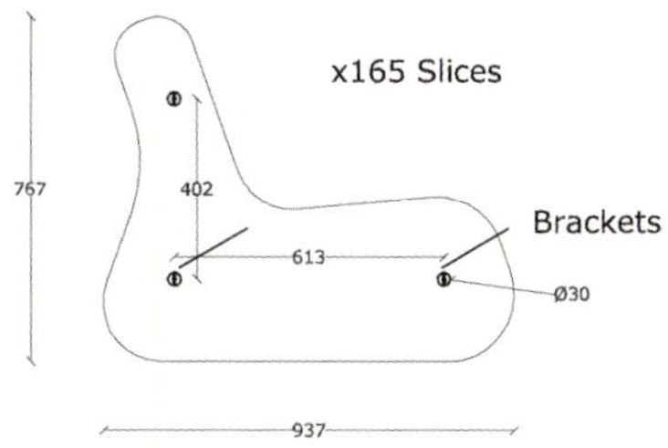

Slice Sc.1-20

Side V. Sc.1-20

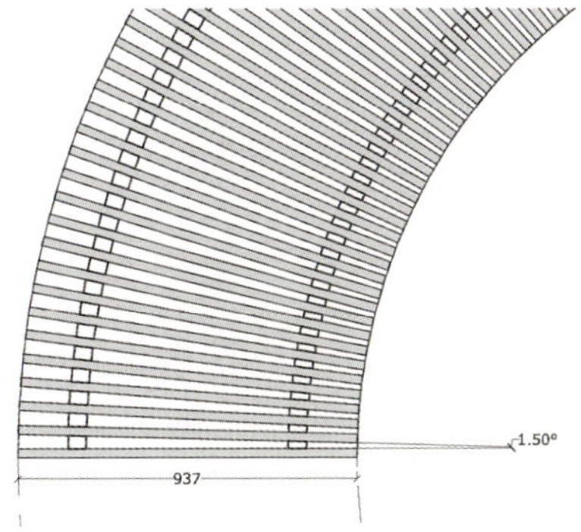

Plan (close-up) Sc.1-20

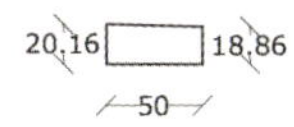

Bracket 1 Sc.1-5

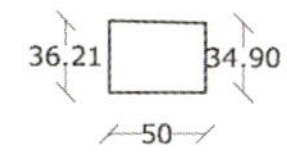

Bracket 2 Sc.1-5

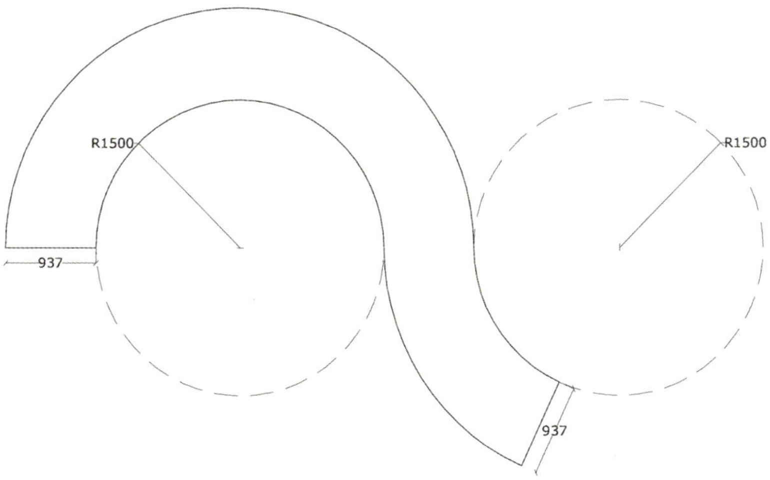

Plan Sc.1-50

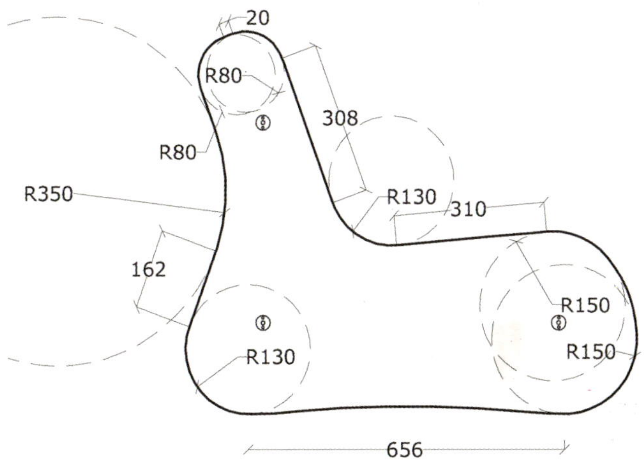

Slice Sc.1-10

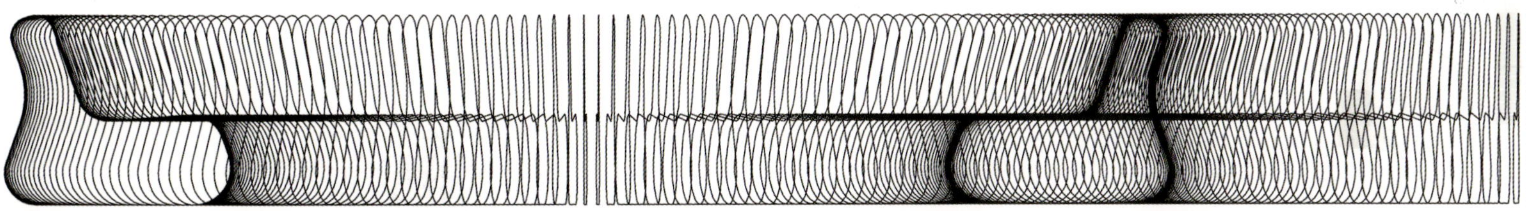

zero emission

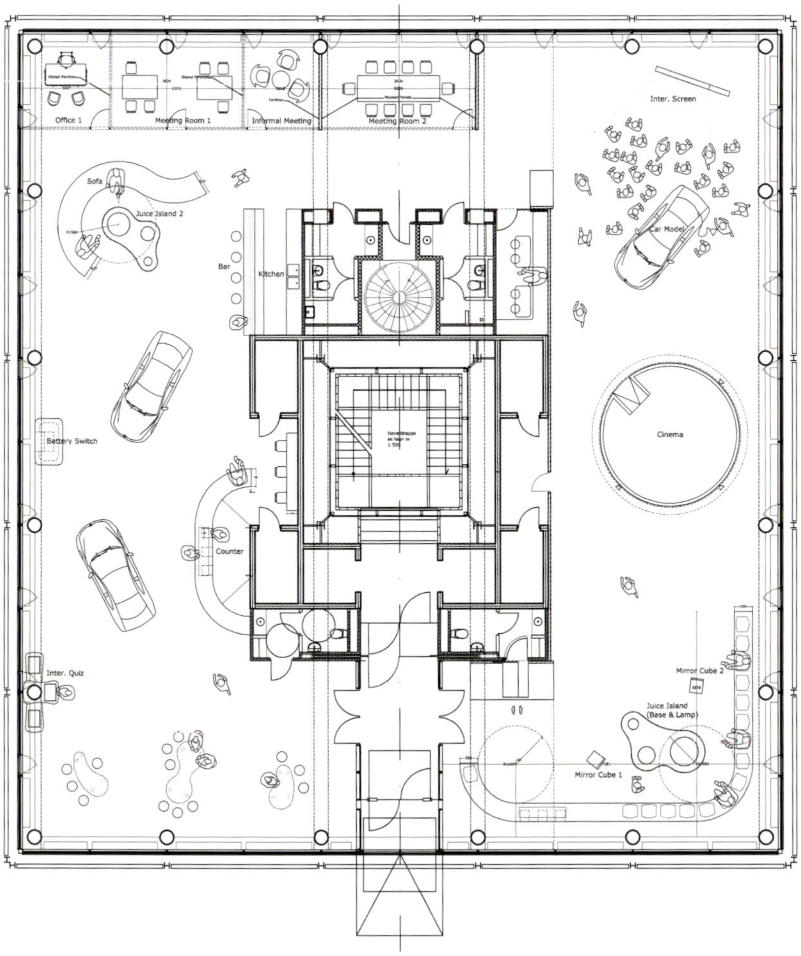
Office 1
Meeting Room 1
Informal Meeting
Meeting Room 2
Inter. Screen
Sofa
Juice Island 2
Bar
Kitchen
Car Model
Cinema
Battery Switch
Counter
Inter. Quiz
Mirror Cube 2
Juice Island (Base & Lamp)
Mirror Cube 1

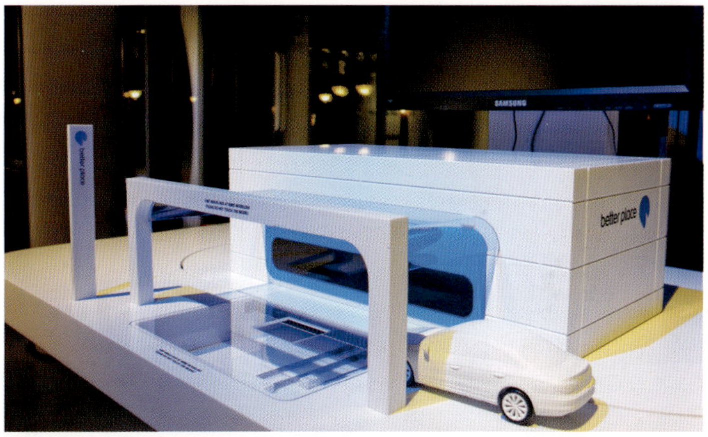
better place

100% elektrisk
med ubegrænset
rækkevidde
better place
DG 24 721

G CLINIC 8F

At first, G Clinic was hair removal clinic for women. When hair restoration medical treatment was also begun, male patients came to visit the clinic. A situation where male and female patients had to wait together in a small 2m^2 room was a serious problem. A doctor decided to move the clinic from the center of Ginza to the edge of Ginza, Shinpashi and Shiodome where high-rise buildings are standing. He demanded luxurious design like in a hotel lounge. The designers use MDF panel because of its cost performance, good workability and dimension accuracy. Although the texture of MDF is somehow cheap, Takafumi Inoue, a furniture fabricator, proposed to utilize oil wax and urethane paint finishing for luxuriously textured surface. 3-dimensional surface is expanded from wall to furniture and the reception counter. From the walls, leafy shade is recreated on the ceiling. We realized a comfortable, open, closed, and continuous space for both patients.

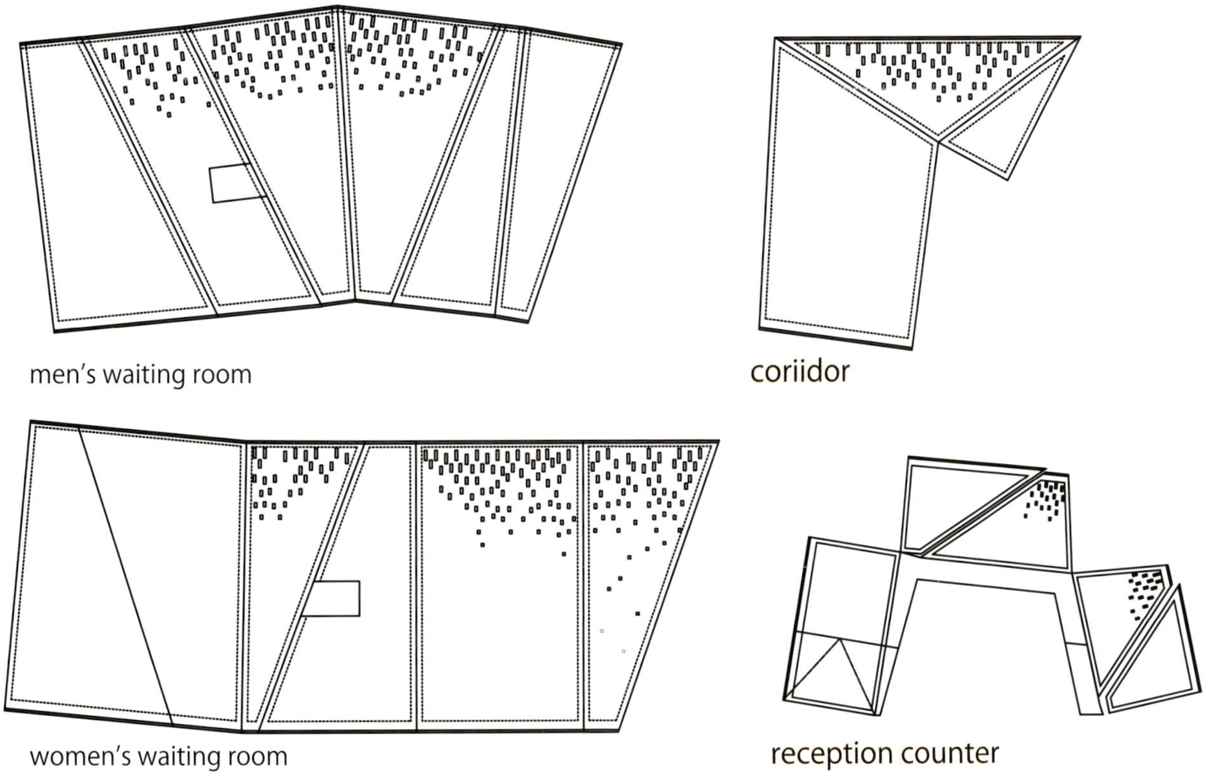

developed figure S=1:100

0.5m 1m 2m

Designer
Maya Nishikori (KORI architecture office) + Yushiro Arimoto (Arimoto Architects Office)

Construction
Takahide Okamoto Yuki Nakazawa(NOBLE)

Furniture Fabrication
Takafumi Inoue, Hideki Yamasaki (Inoue Industries)

Curtain Design
Yoko Ando (Yoko Ando Design)

Curtain Production
NUNO

Client
Dr. Kentaro Masaki (Ginza General Beauty Clinic)

Location
Tokyo, Japan

Built Area
106.83 m^2

Lighting Designer
Izumi Okayasu

Photographer
Hiroshi Ueda

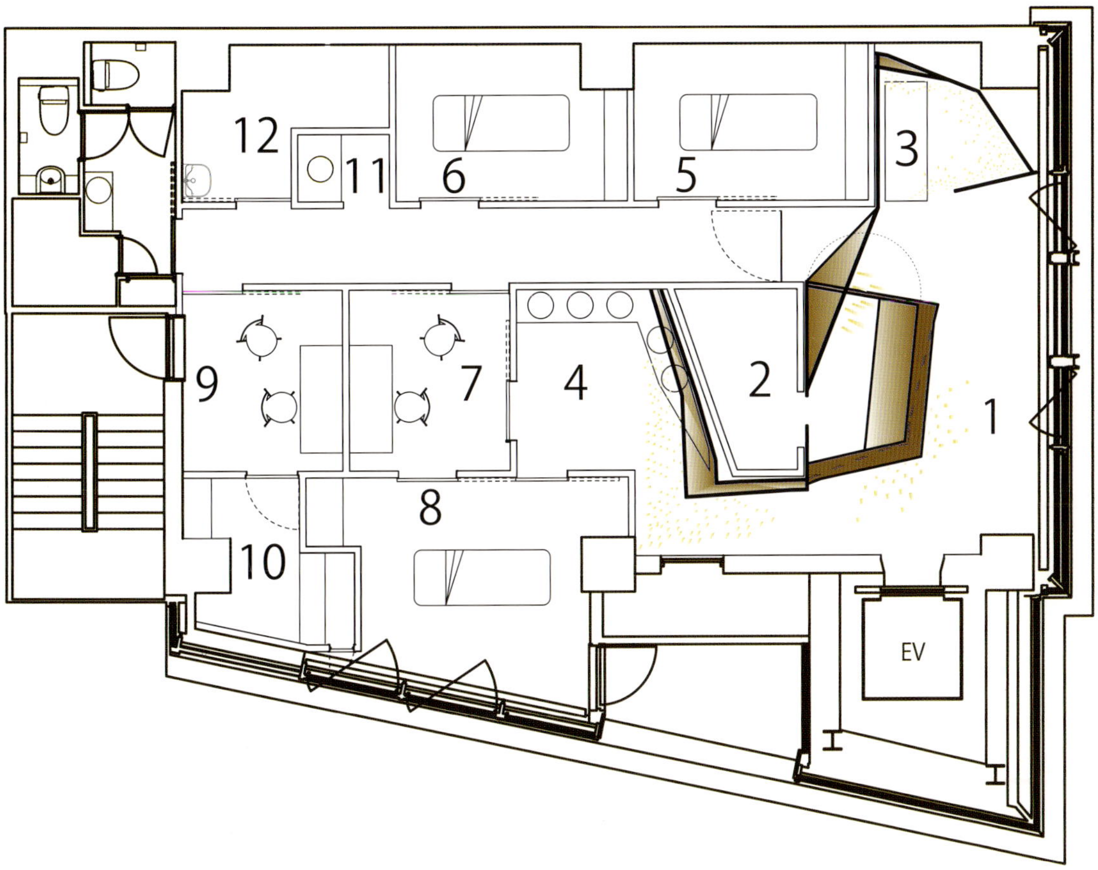

plan

1 entrance space・reception
2 medical record room
3 women's waiting room
4 men's waiting room
5 women's treatment room 1
6 women's treatment room 2
7 consulting room
8 men's treatment room
9 hospital director's room
10 prescription laboratory
11 powder room
12 staff room

0.5m 1m 2m

WAR HORSE: FACT & FICTION

War Horse: Fact & Fiction is a 480 m^2 exhibition aimed at a family audience, which tells the incredible real-life story of the use of horses in warfare right through British military history, tying in with the original War Horse novel by Michael Morpurgo and the Steven Spielberg-directed War Horse film. It deals with the fascinating facts that lie behind the fiction and, whilst it keeps the main focus on the story's World War One setting, the exhibition has a broader storyline and additionally traces the story of horses in war right back to medieval times.

The key objective of the exhibition was to create a dramatic, interactive and immersive experience for the visitor about the story of horses in war, via a series of interactive exhibits and areas, as well as through focus points that reflect the power and impact of the "War Horse" story. Objects from the Museum's archive have been interpreted and displayed in a way that brought them to life to promote a clear contextual understanding. The designers also took care to strike the right balance between creating an entertaining and enjoyable experience and not underplaying the sometimes harrowing notes from history, including the incredible statistic that over eight million horses died in the course of World War One.

Design Agency
MET Studio

Client
The National Army Museum

Location
London, UK

Photographer
James McCauley

Sport & Training
Sport & Training

Requisition, Transport and Training

The Battle of Waterloo
The Battle of Aliwal
Charge!

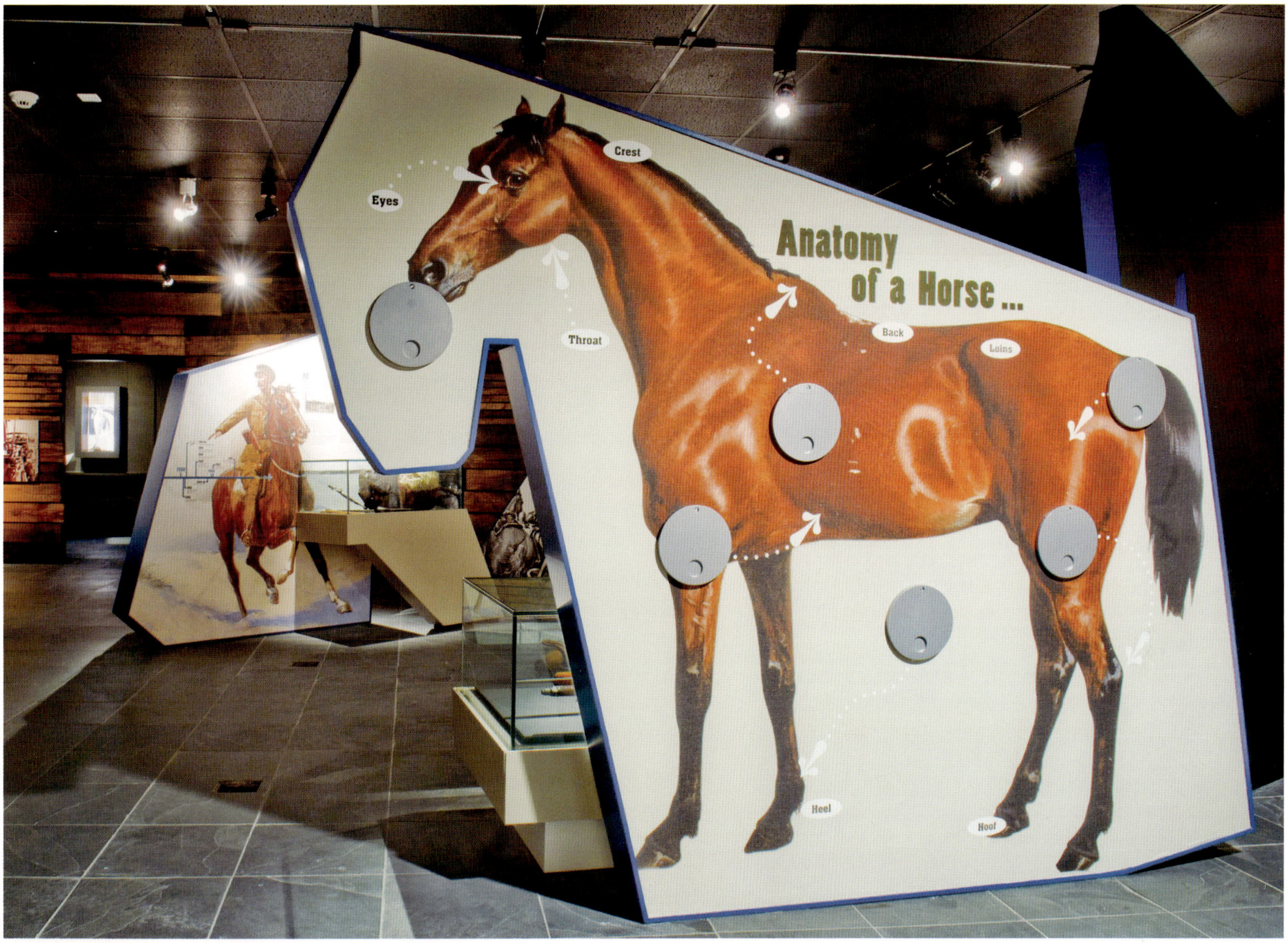
Anatomy of a Horse ...
Eyes
Crest
Throat
Back
Loins
Heel
Hoof

The Battle of Waterloo
The Battle of Aliwal
The Battle of Balaclava

The First World War

The Battle of Hastings

Horses Through Time
The Battle of Hastings
The Battle of Agincourt
'riderless horses careering about wildly in their terror, came plunging into the ranks from the side or in head-on collision'
The Boer War

Riding
Anatomy of a Horse

The National Theatre Production of
War Horse

This was what the soldiers
called 'no man's land'.

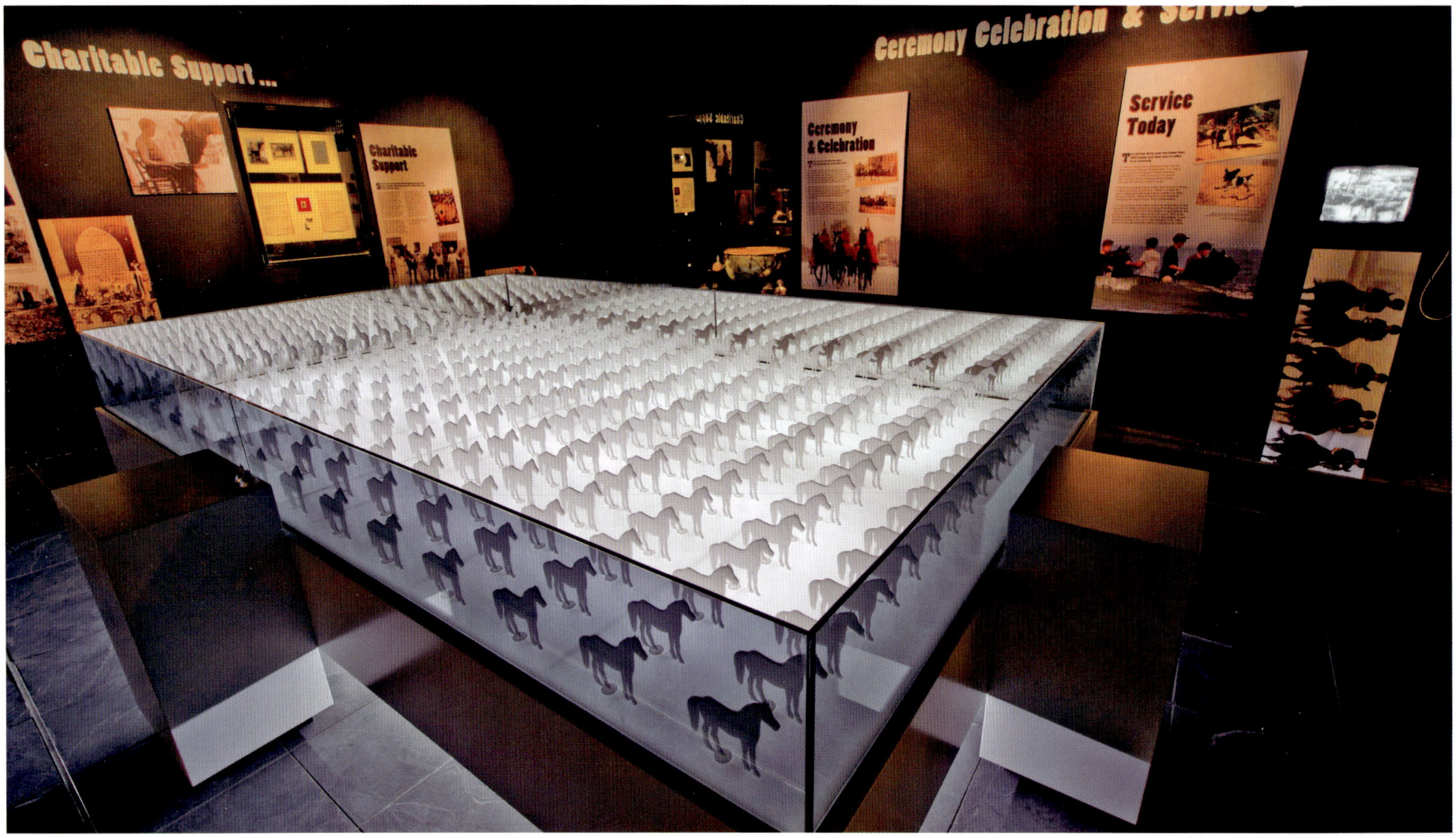
Charitable Support ...
Charitable Support
Ceremony Celebration &
Ceremony & Celebration
Service Today

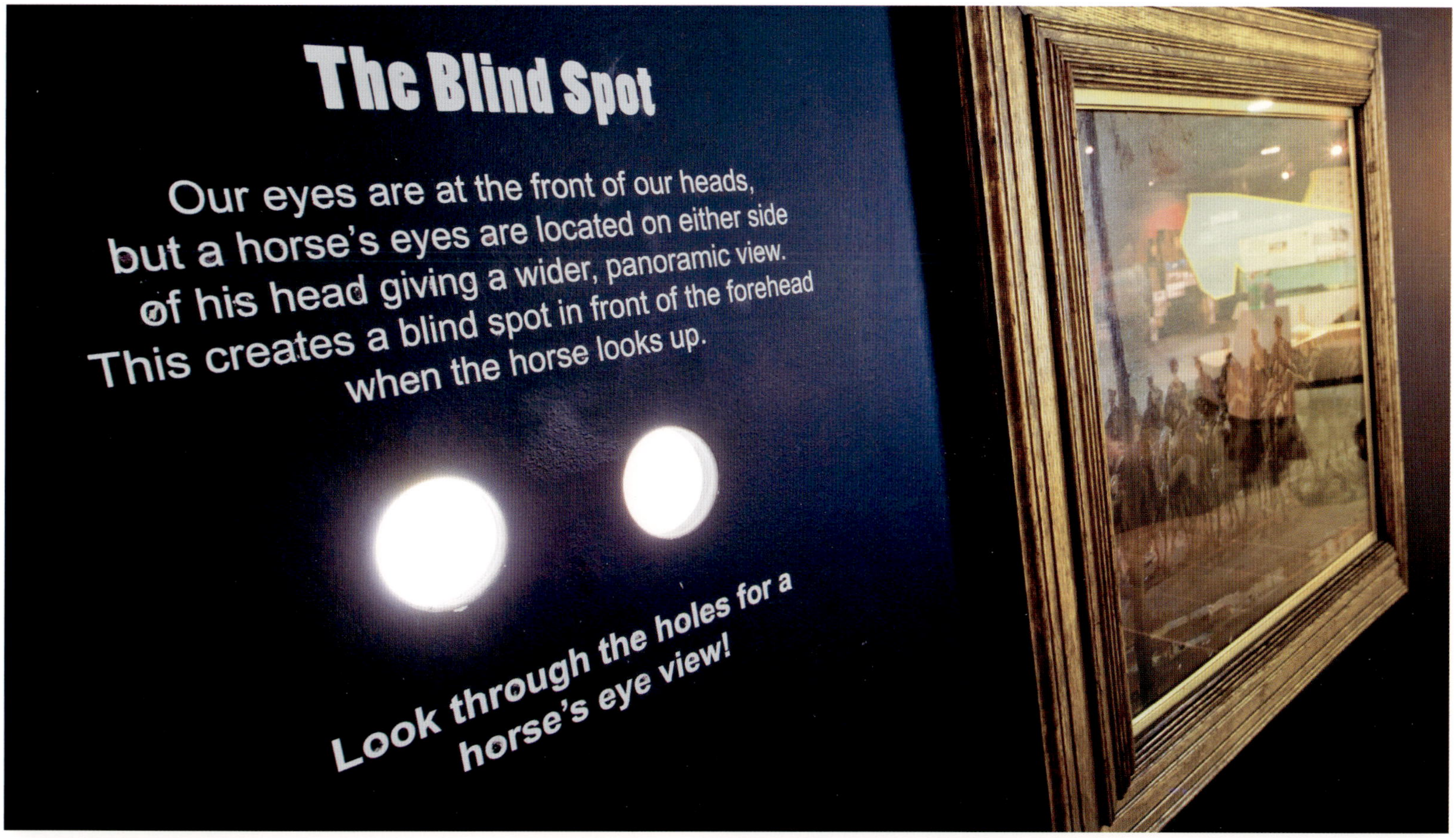
The Blind Spot
Our eyes are at the front of our heads, but a horse's eyes are located on either side of his head giving a wider, panoramic view. This creates a blind spot in front of the forehead when the horse looks up.
Look through the holes for a horse's eye view!

War Horse on Film ...
SEPARATED BY WAR.
TESTED BY BATTLE.
BOUND BY FRIENDSHIP.
WAR HORSE
IN CINEMAS SOON
War Horse The Film
To find out more about Steven Spielberg's War Horse feature film, touch one of the options below ...
Photos from the Film
Making of the Film
Film Trailer

TO BE OR NOT TO BE!

Environmental and equipment design project developed for the new science library in the Pavilion of Knowledge. The intention was to have a sense of "discovery" when entering this room, and to find the books when people have to walk along the space. People deliberatedly wanted to conceal the books in order to avoid the normal bookshelf image of libraries. The contrast between the black on one side and the white on the other, emphasizes the notion of being or not being. Two words are placed on the white surfaces of the shelves, "library" and "investigate", and encyclopedia definitions, such as "to question", "to know", etc. in Portuguese and English. We also designed the tables that are made in lacquered aluminum.

Design Director
Nuno Gusmão

Designer
Giuseppe Greco, Vanda Mota, Joana Proserpio

Client
Ciência Viva

Location
Knowledge Pavilion, Lisbon, Portugal

Photographer
João Morgado

to investigate,
in·ves·ti·gate [in–ves–ti–geyt]
verb (used with object) to examine, study, or inquire into systematically; search or examine into the particulars of; examine in detail;to search out and examine the particulars of in an attempt to learn the facts about something hidden, unique, or complex, especially in an attempt to find a motive, cause, or culprit: The police are investigating the murder.

MEMENTO

Designer
Wesley Meuris

Location
Borgloon, Belgium

Photographer
Kristof Vrancken

The pure, white steel round shape contrasts with the linear patterns of the burial ground. The work's measurements (5 m high and 10 m wide) make it a clearly visible beacon in the sloping landscape. Because Meuris has embedded his work at the rear of the central axis of the terrain, the work does not impose itself. Visitors to the burial site are invited to move towards it. Two subtle openings turning away from the axis allow one to step into the work. At the same time open and closed, the monument is experienced as a space for seclusion, rumination and reflection. A variety of elements enforce the invitation to leave the real world behind. The cylinder shape creates disorienting acoustics. Inside the work, it is quite impossible to make out the sounds penetrating from the outside. Meuris has furnished the inside of the work with a visual pattern of hovering steel lamellae, which come loose from the base wall. Fashioned in high glossy material, these small discs contrast with the flat white base wall, generating an interesting play of shadows and reflection. In terms of shape, the lamellae refer to name plates as a metaphor for the numerous deceased individuals. Because Meuris leaves them blank, one is drawn in mostly by the repetitive structure that fails to provide a (focal) anchor point. Thus, the horizontal path, as one moves towards the monument, takes a vertical flight upwards the moment one enters the monument - the repetitive structure of the cassettes leads the gaze upwards, where there is no ceiling as such to speak of, only sky, a contemporary-style vaulted ceiling, spanning the great divide between what is here and there, the finite and the infinite. As such, the work provides an answer to the commissioners' initial question to create a space for peace and rumination, not bound to any religious denomination.

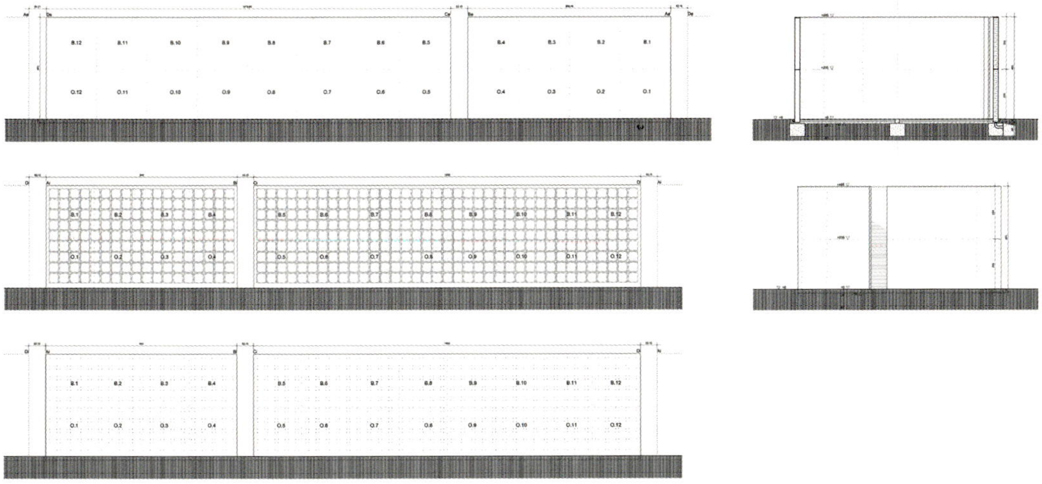

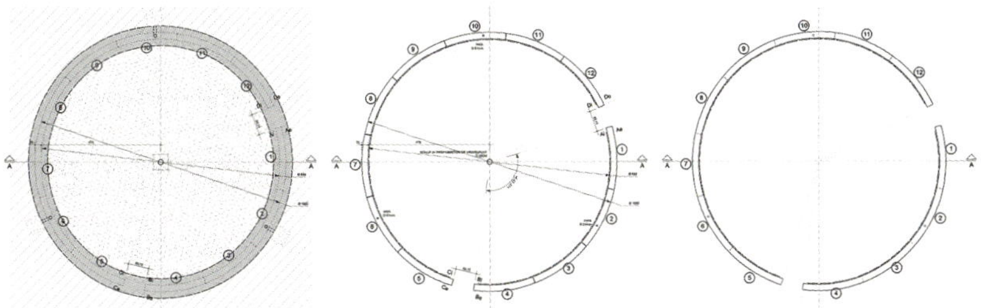

Architect
Carlos Arroyo Arquitectos, Madrid

Associate Architect
ELD partnership, Antwerp

Client
City of Dilbeek

Location
Dilbeek, Belgium

Area
3,554.76 m^2

Photographer
Miguel de Guzmán

ACADEMIE MWD

From certain angles this performance centre in Belgium has a colourful stripy facade, but from others it appears camouflaged amongst the surrounding trees.

Designed by Spanish architect Carlos Arroyo, the Academie MWD is a school of music, theatre and dance at the Westrand Cultural Centre, which is located within a suburban neighbourhood in Dilbeek, outside Brussels.

The architect wanted to come up with a design that mediated between the Westrand building to the west, gabled houses to the east and woodland to the north. "The question was how to harmonize the different situations, and at the same time produce a building with a quality of its own." he said.

Arroyo added a system of louvres to the facade so that, like a lenticular image, the appearance differs depending on the viewing angle.

When facing north-east visitors see a life-size image of trees but when facing south-west they see a mixture of blues and greys that capture the colours of the adjacent building, designed in the 1960s by Belgian architect and painter Alfons Hoppenbrouwers.

Viewing the building straight-on reveals yet another image; a spectrum of colourful stripes and rectangles that are derived from one of Hoppenbrouwers' paintings, while the rear of the building is clad in a similar variation of metal panels with contrasting finishes.

To reference the nearby houses, the massing of the building is broken up into an irregular series of gables. "The new building is a soft transition between the scale of the houses and the imposing presence of CC Westrand." said Arroyo.

The entrance is located beneath a chunky cantilever that contains the main auditorium and theatre, while studios, practice rooms and classrooms are spread across two floors at the other end of the building.

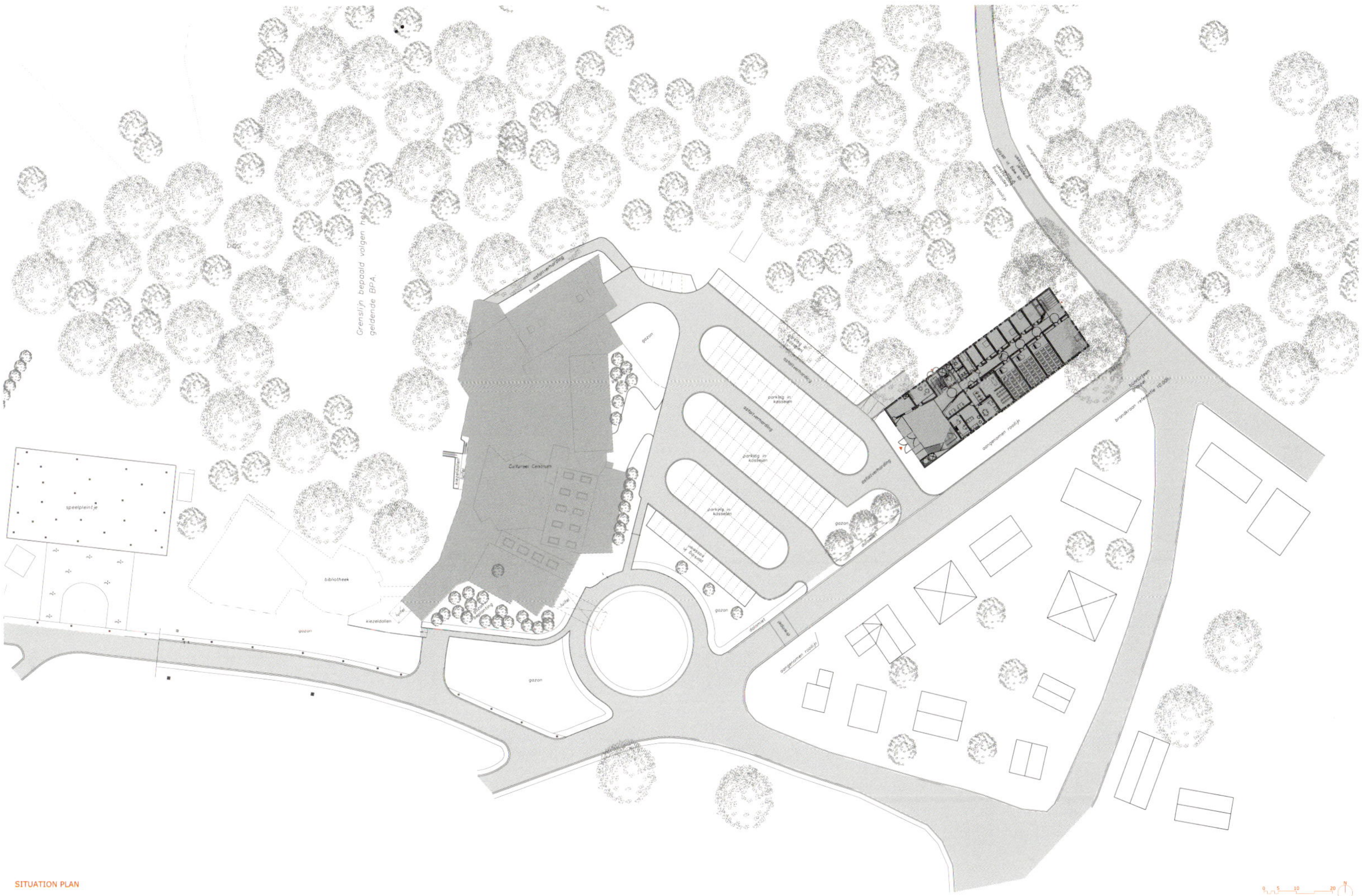

SITUATION PLAN

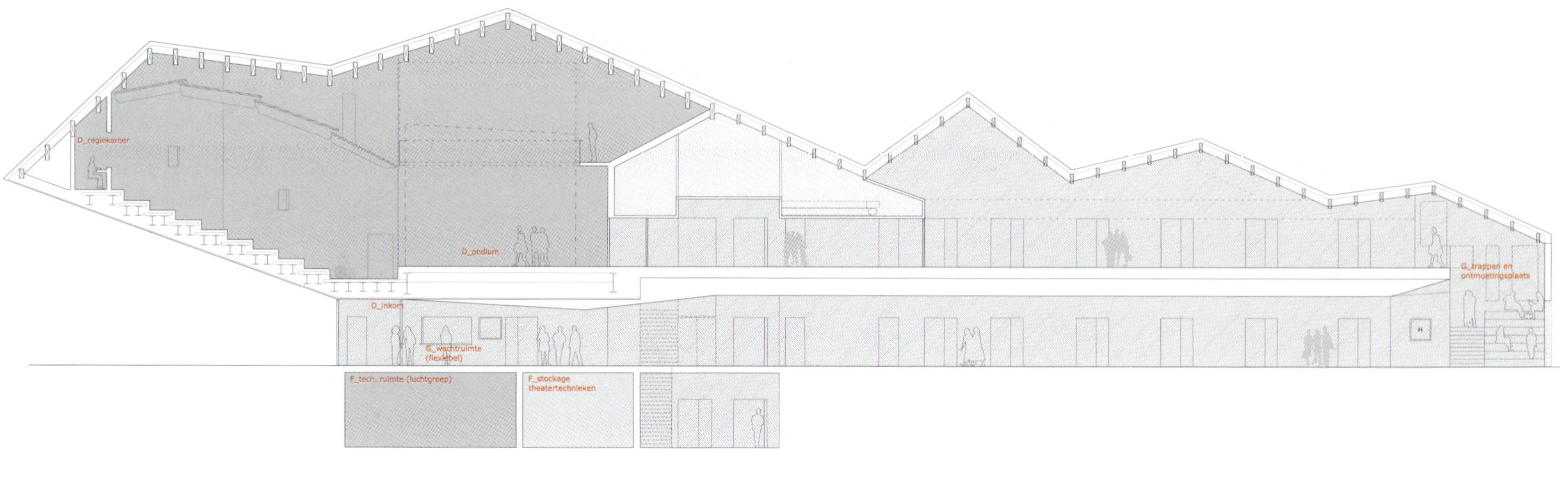
D_regiekamer
D_podium
D_inkom
G_wachtruimte
(flexiebel)
F_tech. ruimte (luchtgroep)
F_stockage
theatertechnieken
G_trappen en
ontmoetingsplaats

Designer
Rosan Bosch

Client
LeasePlan

Location
Brondby, Denmark

Photographer
Kim Wendt

LEASEPLAN

LeasePlan is leading at the Danish car leasing market. Now the company has received a physical design to support its market share. Rosan Bosch Studio has set shape and color on LeasePlan's values and created a design that both puts the customer first and optimizes the workflows for the employees.

In LeasePlan's new showroom an open reception desk offers the customer welcome and stresses the communication: here the costomer is in focus – not the car. The colors of the company's logo characterize the interior and create a beautiful and exclusive framework for the showroom.

Rosan Bosch's design for LeasePlan's new headquarters focuses on the meeting and communication with the customer as well as the values that characterize the company. This applies, not just to customer facilities, but also to the underlying administrative buildings.

With identity graphics, custom-designed meeting facilities and inclusive public spaces, a bridge between showroom and office areas is created. The office areas convey the same participatory experience for the staff as a showroom does for the customers. In this way, the development and design project gathers the aims of LeasePlan to prioritize the best customer experience while at the same time creating a better workplace for the employees.

STILLERUM
MØDEØ
WORKSHOP
KAFFEØ
TEKØKKEN
PRINT
VAND
LOUNGE
KAFFE
LOUNGE
DUESLAG
RECEPTION
SHOWROOM
HØJBORD
HØJBORD
BAGPROJEKTION
PROJEKTION
GARDIN
PROJEKTION
AUDITORIUM
ROSANBOSCH
STUEPLAN 1:100

LeasePlan

LeasePlan
it's easier to leaseplan

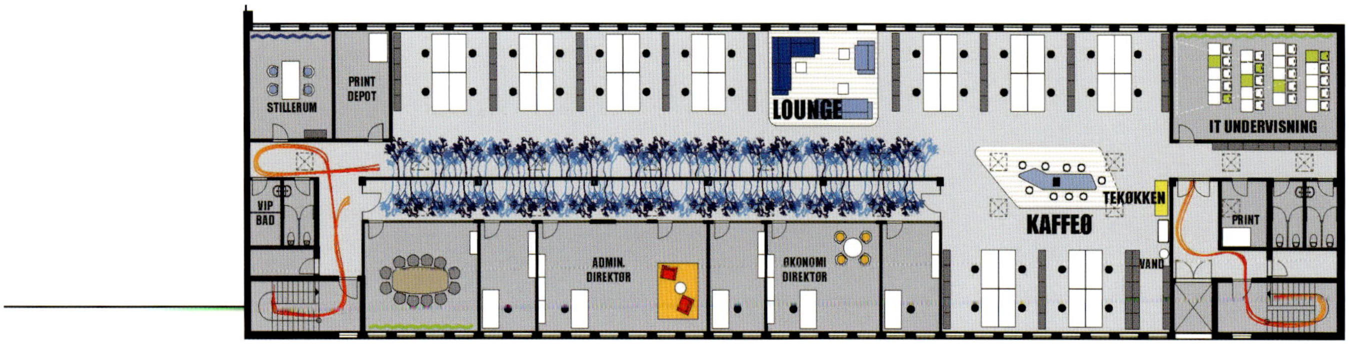

1. SAL PLAN 1:100

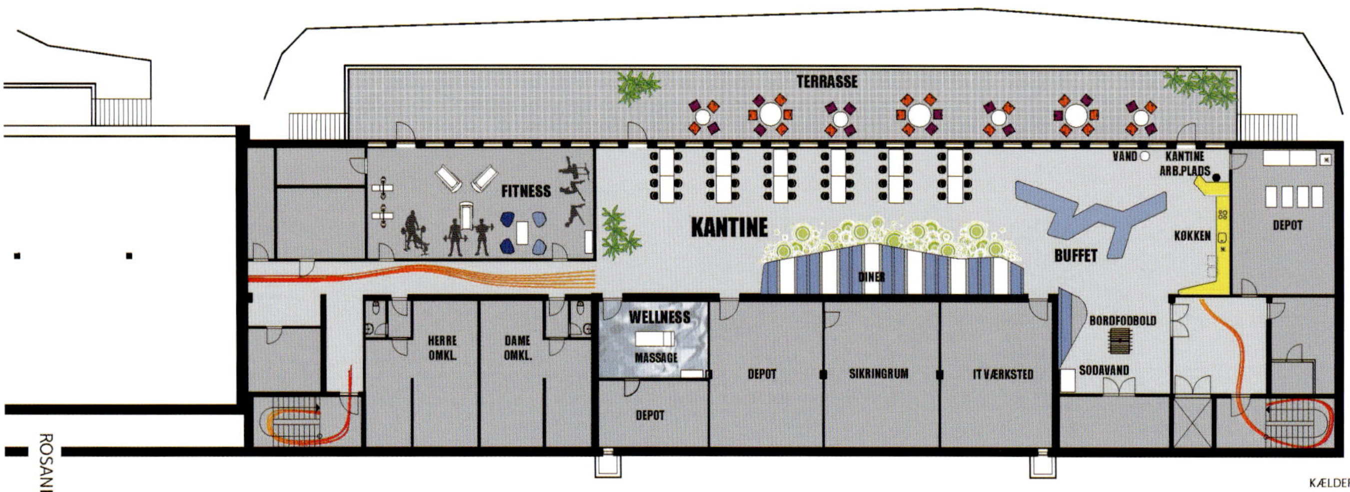

KÆLDER PLAN 1:100

ROSANBOSCH

POWERTEC

STATE OF GREEN

When 50 Danish businesses and industries along with the Danish crown prince couple on November 21-24th, 2011, made for Australia to promote the Danish Business community at the business promotion event "state of Green" in sydney and melbourne, art and design held a special significance as platform for communication of the business event's themes of sustainability and green energy. By artistic tables, heat-sensitive installations and living chandeliers, the art and design studio Rosan Bosch created for the Danish arts agency a series of artistic initiatives for the Danish business promotion event that created a unique communication platform for the themes of the event and contributed to the branding of Denmark as an innovative nation.

Among other things, a series of "conversation pieces" on the tables for the business event's seminars and official lunches replaced the traditional centrepiece with a range of artworks that in a creative and artistic manner interpreted the event's themes of sustainability and green energy and gave the participants new perspectives to talk about. Lunch was therefore not taken in traditional surroundings but instead with a view to artistic installations with flasks filled with toilet detergent and other strange liquids, heat-sensitive sculptures that encouraged to be touched and artworks made of matches that with their well-placed matchboxes tempted the participants to strike the decisive match. At the same time, two artistic lamp installations staged the traditional conference facilities with a creative interpretation of the four elements, wind, water, earth and fire projected in living images on the sculptural chandeliers. In the reception area, a uniquely designed registration table and humoristic and challenging signs welcomed the participants and, among other things, encourages the attending businessmen and other participant to swap neckties.

Together, the different creative initiatives created a completely new type of business promotion event where art, design and culture combined became an important communication platform for the event and its values. With its direct use of creative and artistic elements, "state of Green" was an event where experience and involvement served as important means to create an image of Denmark as an innovative and creative nation.

2
SOLAR & OTHER RENEWABLES
State of Green

Designer
Rosan Bosch

Client
the Danish arts agency

Location
Copenhagen, Denmark

Materials
aluminium, projection screen, projectors, plastic, fans, watermelons, flasks, detergent, ceramic clay, acrylic plates, heat-sensitive fabric, matches, elastic bands, wood, cardboard, straws, white-pigmented ash wood laminate.

Photographer
Palle Demant, Kim Wendt

Imagine a more
sustainable
future
State of Green
REGISTRATION

Refreshments

Think
Green

THE STENO MUSEUM "DARLING BODY, DIFFICULT BODY"

Designer
Rosan Bosch

Client
Steno Museet

Location
København S, Danmark

Area
120 m^2

Photographer
Laura Stamer

Am I too fat or too thin? How tall is it normal to be? How much food do I really need to eat every day? Many young people today are not comfortable with their bodies, and have difficulties in finding out what is up and down and right and wrong in the vast jungle of beauty icons, body ideals, lifestyles and food cultures.

In collaboration with the Steno Museum in Aarhus, the Rosan Bosch studio has created the sensuous and engaging exhibition "Darling Body Difficult Body", focusing on the surrounding's influence on our bodies — an exhibition created with a special focus on young people and their "difficult bodies". The unusual and unique touch to this exhibition is Rosan Bosch's contribution, using her artistic experience and expertise to create an inclusive and participatory communication platform where artistic and creative surroundings and scenarios provide the framework for the exhibition's themes, offering the young people the opportunity to try them on their own body, as for example through a heat sensitive lounge area where you leave an imprint of your naked body and an interactive locker room that creates a focus on taboos and modesty through artistic staging of communication and set-design.

Bruge du din kro
være nd? Bader d
Hvem estemmer o
viden aben? Er du
skrækket
uger d

OMKLÆDNINGSRUMMET
OM AT VÆRE NØGEN
står vi nøgne for at vaske sved og kropslugte af.
For nogle år siden gik alle skoleelever i
er der mange. som ikke gør det.
at acceptere sin egen og andres nøgenhed?

Kære krop, svære krop
Jeg er mig.
Jeg er optaget af min egen krop – og af de andres kroppe.
Men hvad tænker andre om mig? Og hvad skal jeg rette mig efter? Politikerne som siger: "Lev sundt!" Fødevareindustrien som siger: "Spis vores mad!" eller modeverdenen som siger: "Køb dine drømme her!"
Eller skal jeg bare lytte til mig selv?
Kroppen vil følge mig hele livet og jeg skal passe godt på den.
Men hvad er så det rigtige at gøre?....
UDSTILLINGEN HAR FÅET STØTTE FRA
TAK FOR ET GODT SAMARBEJDE

1970 - 80'erne
Bodybuiling - byg din krop

UNIVERSITY COLLEGE OF NORTHERN DENMARK

University College of Northern Denmark (UCN) aims to be a leading provider of future-oriented education. Rosan Bosch has developed a design that ties the education programmes together and promotes interaction, cooperation and knowledge sharing. With its new interior, UCN now has a learning environment that facilitates new ways of working and learning for students, lecturers and consultants.

Cave-like lounge areas encourage relaxed interactions and focused concentration. Interactive platforms enable knowledge sharing across disciplinary boundaries, and the common areas are incorporated into the course activities, which gives students greater insight into each other's everyday activities. Everything is tied together by graphic elements in wallpaper designs and collages that act as identity markers for the individual programmes.

The new interior design at UCN was developed with the goal of facilitating cooperation and communication as key aspects of learning. at the Centre for Educational ressources, the new design serves as a tool to rethink the centre's professional competencies with new means for communication and interaction. In the centre of the building one finds the "Planet", a huge digital sculpture that serves as a communication platform connecting the many education programmes offered at UCN. The sculpture serves both an involving and an informative purpose and links the students together across disciplinary boundaries.

The new interior has given UCN a distinct profile with inviting common areas that highlight both internal and external activities. The spatial design also promotes the educational principles and serves as an active instrument of learning and development in everyday activities.

Designer
Rosan Bosch

Client
University College of Northern Denmark

Location
Hjørring, Denmark

Photographer
Kim Wendt

NYHEDE
CFU
cooperative learning
Cooperative Learning
Teambuilding
Classbuilding

HISTORIE
BIOLOGI DANSK 0.-2.KL.
DANSK FRA 3.KL. DANSK
SOM ANDETSPROG DAT.
DRAMA ENGELSK
BILLEDKUNST FRANSK
FYSIK/KEMI GEOGRAFI
HISTORIE
HJEMKUNDSKAB
HÅNDARBEJDE IDRÆT
ANDRE SPROG
MATEMATIK MEDIE
MUSIK PSYKOLOGI
RELIGION
NATUR/TEKNIK
SAMFUNDSFAG
SUNDHED TVÆRFAGLIG
TYSK SLØJD
ERHVERVSØKONOMI
UDDANNELSE ØVRIGE
FAG FRILÆSNING
PÆDAGOGISK ALMEN
PÆDAGOGISK UDLÅN
FAGLITTERÆR UDLÅN
SKØNLITTERÆR
SPECIALPÆDAGOGIK
PÆDAGOGISK DANSK
MEDIEKUNDSKAB
FYSIK/KEMI DANSK
GYMNASIE NYHEDER
PÆDAGOGISK
INDSKOLING

Designer
Rosan Bosch

Client
Vittra Brotorp

Location
Brotorp, Sweden

Photographer
Kim Wendt

VITTRA SCHOOL BROTORP

Rosan Bosch Studio has created a colorful and imaginative interior for the newly built Vittra School in Brotorp. The design supports the school's pedagogical methods and its emphasis on offering students and teachers diverse environments depending on the learning situation. The new design functions as an important tool in the schools' daily life.

Vittra Brotorp has been equipped with flexible learning spaces, inviting gathering places and small niches for concentration and contemplation. The interior design gives teachers the opportunity to provide information to large groups after which the students can work on assignments individually or in small groups in the customized interior. In that way the design leads the school's educational ideals into practice.

Vittra Brotorp has custom designed sections for three different age groups — a custom designed library and a multi-colored podium are among the conspicuous design elements. Rosan Bosch Studio has designed a common area for the kindergarten children where a green structure bays through the space in various shapes and it forms a landscape for play and exploration. The structure is built in children's height and they will feel comfort in the small pockets and niches – while at the same time adults can create an overview of the entire room.

Vittra School Brotorp is part of the Swedish free school organization Vittra. Rosan Bosch Studio has also developed interior designs for Vittra Telefonplan and Vittra Södermalm.

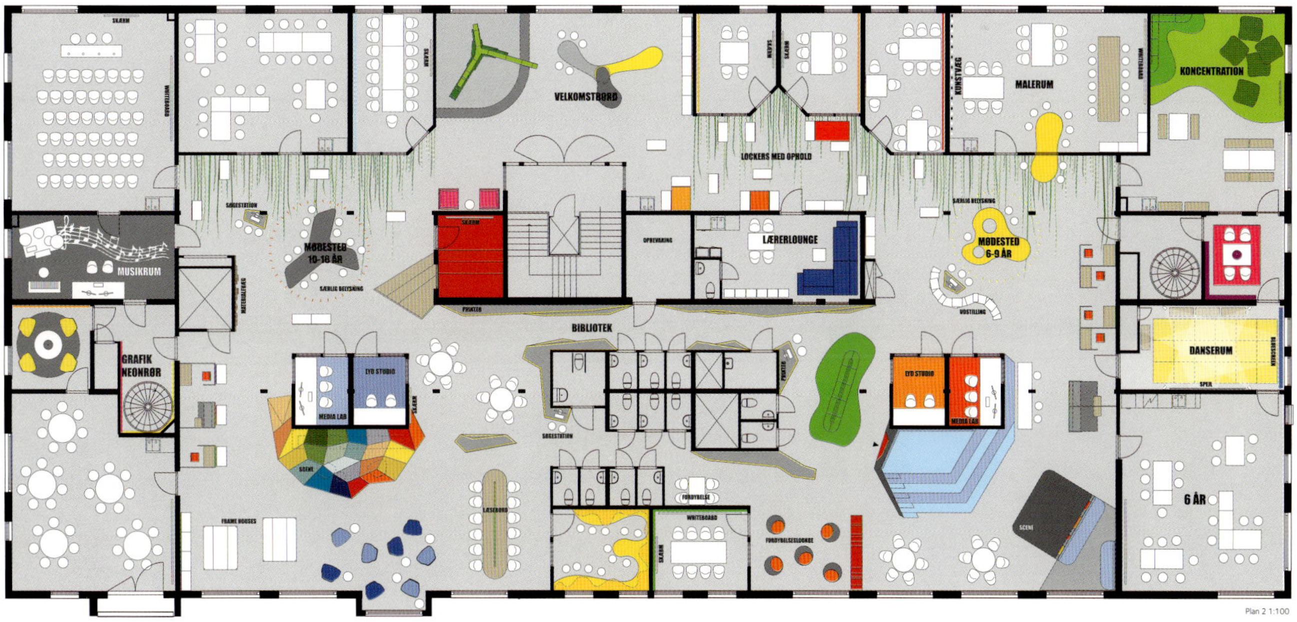

Plan 2 1:100

Plan 1 1:100

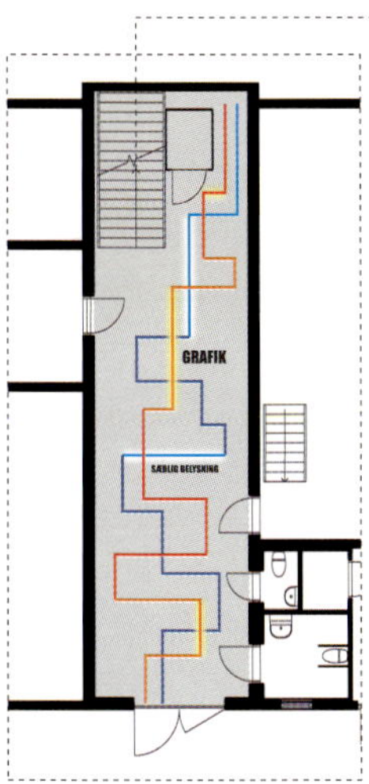

Plan 0 1:100

ROSANBOSCH

Strukturen set skråt fra siden

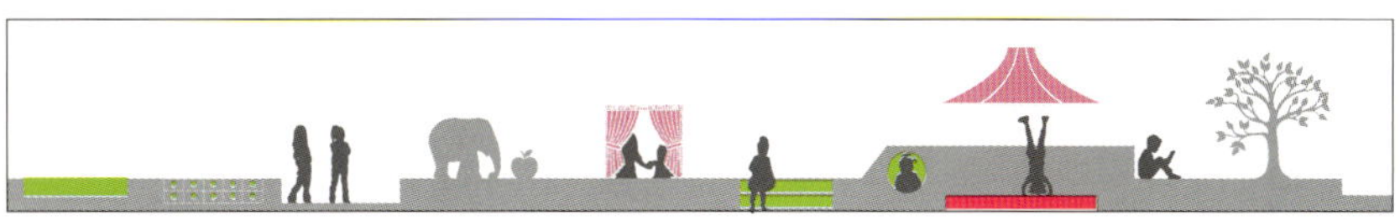
Diagramatisk opstalt

Vittra skole Brotorp

23 april 2012

VITTRA SCHOOL SÖDERMALM

Designer
Rosan Bosch

Client
Vittra Södermalm

Location
Södermalm, Sweden

Photographer
Kim Wendt

Crystal chandeliers, colorful caves and a library that opens like a treasure chest. Rosan Bosch Studio has transformed common areas at the Swedish school Vittra Södermalm into inspiring learning environments that break down the boundary between education and leisure.

Vittra Södermalm has 350 students and is located in a historic building in central Stockholm. The new design strives to support the school's pedagogical methods and gives teachers and students the opportunity to work in different settings depending on the learning situation. The design varies from small caves for concentration and contemplation, a colorful cave with deep red upholstery to organic high tables for group work and soft lounge furniture for informal gatherings. The large reading table brings back memories from American university libraries and gives a sense of learning and seriousness.

"Rosan Bosch has created a learning environment that helps us keep our educational visions into reality," says Annica Ångell, rector of Vittra Södermalm. "She has created an environment that gives the students the opportunity to seek out different environments depending on their needs. It's a huge support in their daily work."

With the new interior the school's large common area has been the focal point for teaching, tutorials and social activities. In the middle of the area a huge black box with wavy red graphics shapes an unconventional framework for the school library. The library contains of books, magazines, ipads, laptops, working material and it lures passers with its colors and built-in light in the shelves.

"I see it as our job to create an environment where students thrive and feel happy being at school," says Rosan Bosch. "I have tried to break down the boundary between leisure and work by building this "treasure chest" where it's nice to stay while at the same time there are tools for learning and inspiration."

Vittra Södermalm is part of the Swedish free school organization Vittra. Rosan Bosch Studio has also developed interior designs for Vittra Telefonplan and Vittra School Brotorp.

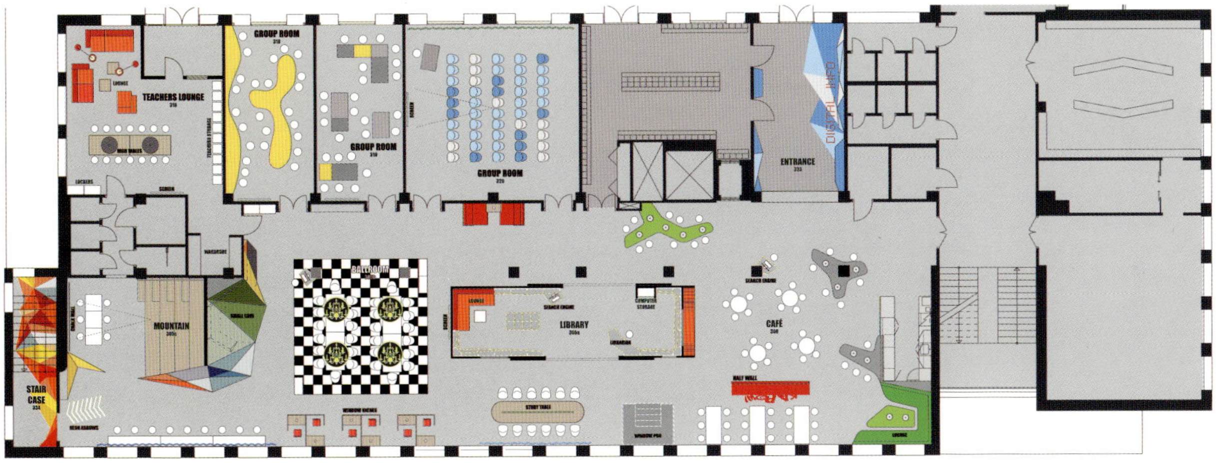

PLAN 2 1:100

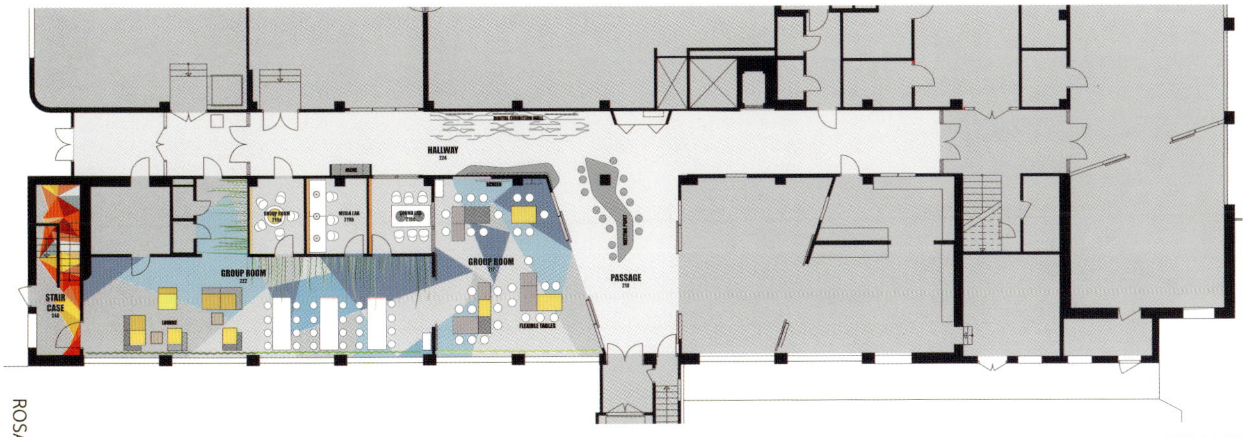

PLAN 3 1:100

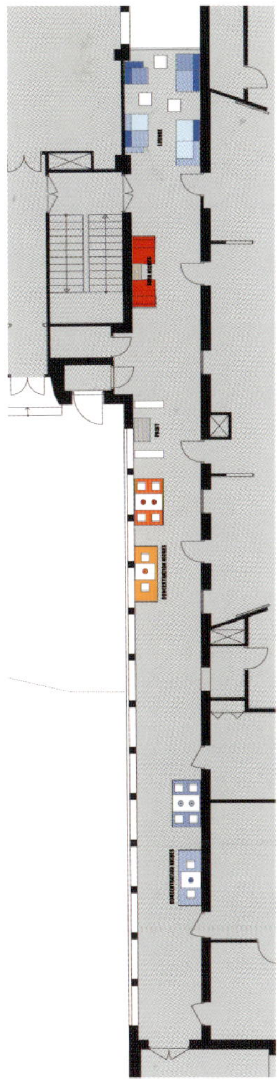

PLAN 3 1:100

VITTRA SCHOOL SÖDERMALM
MAY 9 TH, 2012

VITTRA TELEFONPLAN

Instead of a classic classroom setup with desks and chairs, a giant iceberg with a cinema, a platform and room for relaxation and recreation now forms the setting of many different types of learning situations. The Swedish free school organization Vittra's new school in Stockholm has a physical design that promotes the school's educational methods and principles.

When the new Vittra School "Telefonplan" was established in Stockholm, Rosan Bosch created the school's interior design, including space distribution and distinctive custom-designed furnishings. The interior design revolves around Vittra's educational principles and serves as an educational tool for development through everyday activities.

Vittra puts a high priority on developing new methods for teaching and interaction as a basis for educational development. Instead of conventional classroom teaching, the students are taught in groups adjusted to their achievement level based on the school's educational principles about "the watering hole", "the show-off", "the cave", "the campfire" and "the laboratory".

Rosan Bosch has used challenging custom-made furnishings, learning zones and room for the individual student to facilitate differentiated teaching and learning in a school where the physical space is the most important tool for educational development. Instead of a classic desk-and-chair setup, for example, a giant iceberg that features a cinema, a platform and room for relaxation and recreation now forms the setting for many types of learning situations, and flexible labs provide opportunities for focusing on special themes and projects.

The designs and the interior also accommodate Vittra's active efforts to incorporate digital media and digitally based didactics. In the Vittra schools, laptops are the children's most important tool — whether they are working sitting down, reclining or standing up.

After the project was completed, the results were translated into a design manual that will set a precedent for the design of Vittra's other schools in Sweden.

Designer	Client	Location	Area
Rosan Bosch	Vittra AB	Stockholm, Sweden	1,900 m^2

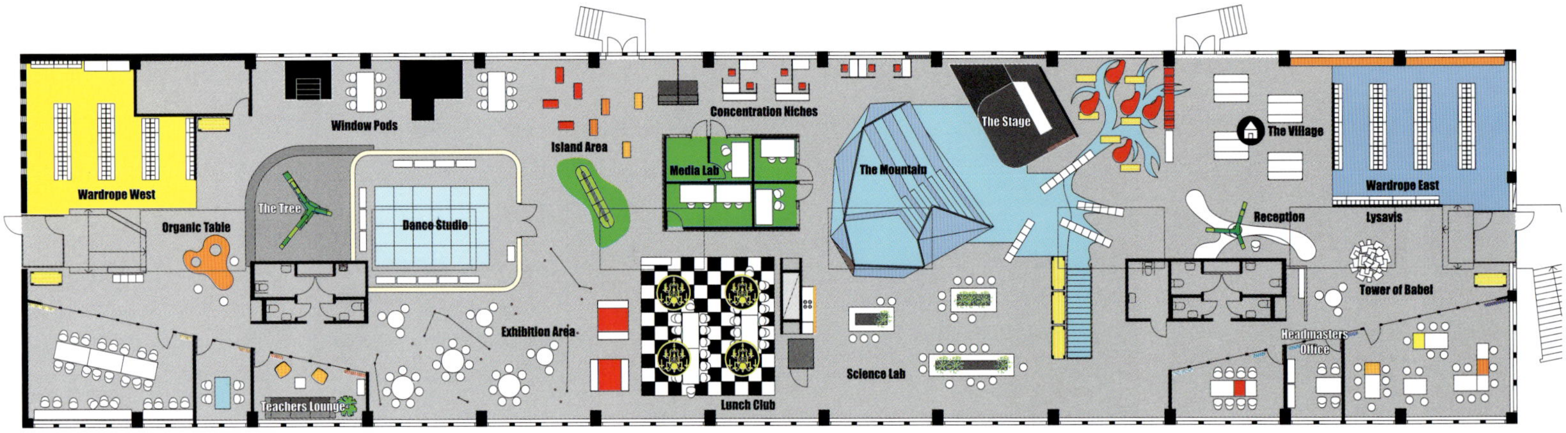

Window Pods
Concentration Niches
Island Area
The Stage
The Village
Wardrope West
The Tree
Dance Studio
Media Lab
The Mountain
Wardrope East
Organic Table
Reception
Lysavis
Tower of Babel
Exhibition Area
Headmasters Office
Science Lab
Teachers Lounge
Lunch Club

Workshop
Canteen Area

MONI NILSSON
SEMLAN OCH GORDON
KÄRLEK KYSSAR OCH REKORD
Sago prinsessan
Dinosaurier

SCHIPHOL LOUNGE 4

Our second collaboration with Schiphol Airport has resulted in the redesign of Lounge 4. The project consists of the design of two retail sections and a waiting area. The starting point of the project was to convey the feeling of a cozy neighborhood rather than a huge international airport. This resulted in the suggestion of a shopping street composed of a variety of buildings. For the seating area we integrated Joep van Lieshout's sofa design for Lensvelt, spreading out like a beautiful fresh green grass landscape. Two prominent stands offer differences in level for people like to have an overview. And kids will find a wonderful integrated slide. Because the lounge does not have a view on planes we added a couple of specially designed plane benches. For does know what it feels like to desperately seek a socket to power up your mobile device at an airport the designers have integrated more than 100 power sockets in the scheme! And finally the original ribbon like M.C. Escher artwork integrates beautifully in the esthaetics of the lounge.

Builder
Hekker, Lensvelt

Client
Schiphol

Area
2,000 m^2

Design Team
Frank Tjepkema, Leonie Janssen

Location
Schiphol Airport, Amsterdam

Photographer
Mike Bink

Gates M
Toilets

Gates M
Toilets

Tjep.
congratulates Schiphol
with the opening
of lounge 4
tjep.com
O·P·I

Lounge 4
TOMMY HILFIGER

JOH 3 – RESIDENTIAL BUILDING

Property development group Euroboden is realizing a unique residential building at Johannisstraße in Mitte, Berlin's downtown district. J. MAYER H. architects' design for the building, which will soon neighbour both Museum Island and Friedrichstrasse, reinterprets the classic Berliner residential building with its multi-unit structure and green interior courtyard. The sculptural design of the suspended slat facade draws on the notion of landscape in the city, a quality visible in the graduated courtyard garden and the building's silhouette and layout. Plans for the ground floor facing the street also include a number of commercial spaces. The generously sized apartments will face south-west, opening themselves to a view of the calm, carefully designed courtyard garden. Spacious, breezy transitions to the outside create an open residential experience in the middle of the city that, thanks to the variable heights of the different building levels, also offers an interesting succession of rooms. The units' varying floor plans and layouts indicate a number of housing options; condominiums are organized into townhouses with private gardens, classic apartments or penthouses with a spectacular view of the old Friedrichstadt. The integrated design concept, which incorporates everything from façade to stairwells, elevators to apartment interiors, promises a unique spatial and living experience with an eye to high design.

SITEPLAN M. 1:1000

Scale 1:500
1st Floor

Scale 1:500
Ground Floor

Design Agency
J. MAYER H. Architects

Project Architect
Hans Schneider

Project Team
Juergen Mayer H., Marcus Blum, Wilko Hoffmann, Filipa Frois Almeida

Competition Team
Juergen Mayer H., Thorsten Blatter, Marcus Blum

Architect on Site
Architekturbuero Wiesler, Stuttgart with Thomas Quinten Projektmanagement, Berlin

Client
Euroboden Berlin GmbH

Location
Berlin, Germany

Structural Engineer
EiSat GmbH, Berlin

Building Services
Ingenieurgesellschaft Striewisch mbH

Building Physics
Ingenieurbüro Santer, Duisburg

Fire Security Consultant
Fire Safety Consult, Berlin and KLW Ingenieure GmbH, Berlin

Photographer
Ludger Paffrath and Patricia Parinejad for Euroboden, Rick Jannack

PAPALOTE VERDE MONTERREY

It is an underground museum which integrates state of the art construction and sustainability technologies to the building with an interactive and dynamic approach.

Scheduled to open in winter 2012, the Children's Museum "Papalote Verde Monterrey" is expected to become Mexico's first LEED Platinum certificate.

In this project the concept of "museum" is addressed from an entirely new perspective, as the building becomes part of the learning experience. The flowing geometry of the building, jointly with its demonstrative technologies, teaches visitors to incorporate passive and active systems to the construction, thus allowing museography and architecture to merge.

Usually a museum is conceived as a series of rooms that explain a topic and follow a rigid sequence. In this case, the topic is being explored throughout the museum as a whole and understood within diverse contexts, allowing the visitor to create his own tour.

The project explores two main strategies; one, the restoration of two existing buildings on site and two, the underground construction of the new building. The latter allows for a third strategy to be emerged, the landscape proposal, allowing the new construction to blend into the existing morphology of the site. The restoration of the two existing buildings allows us to introduce a new program to an under-used space and improve its current condition. The underground strategy reduces the impact on the site and helps maintain the park's horizontal landscape.

Design Agency
Iñaki Echeverria

Client
Papalote Museo del Niño (Children's Museum)

Location
Monterrey, Nuevo Leon, Mexico

Site Area
9,908 m^2

Project Area
15,920 m^2

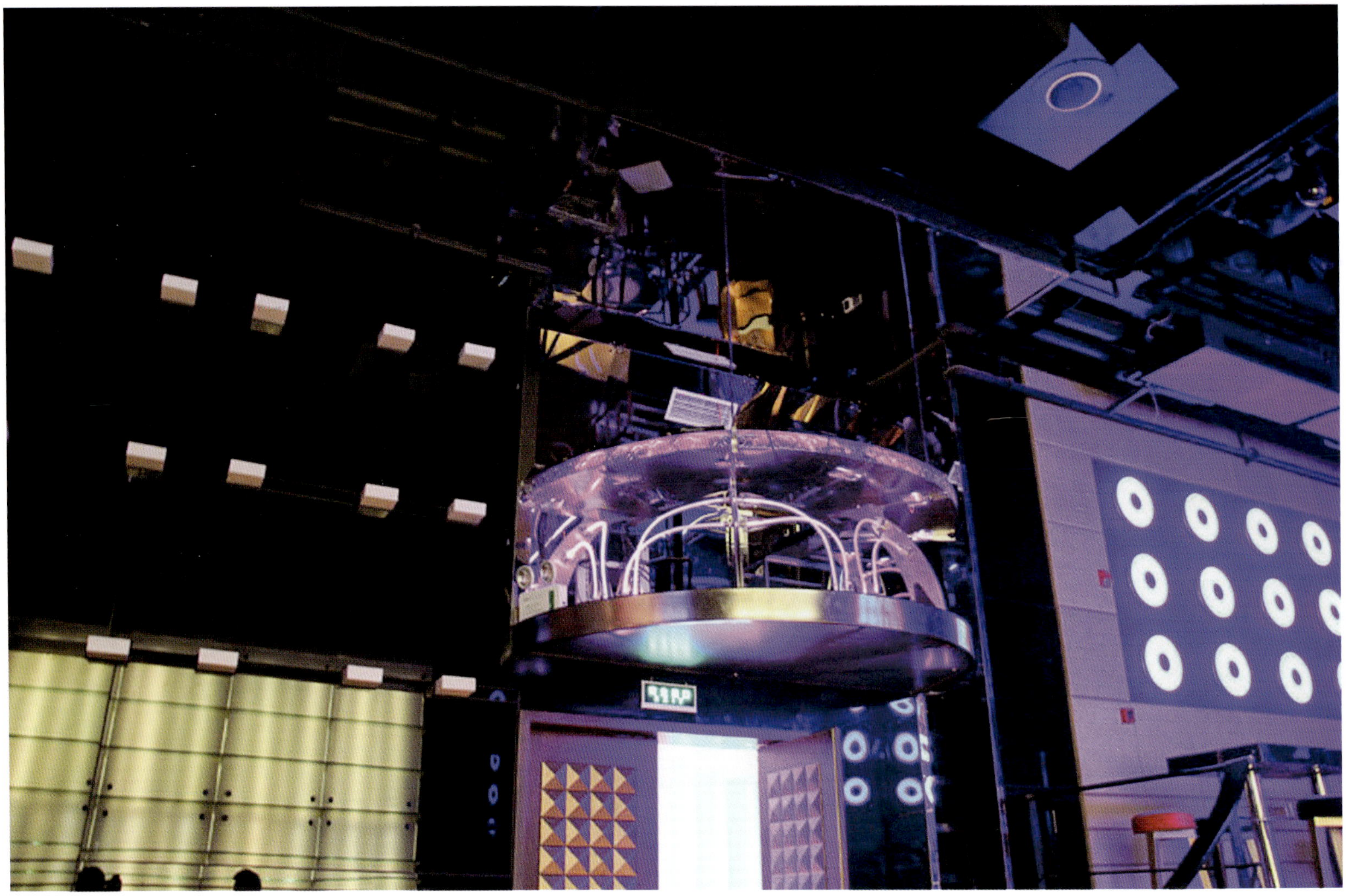

ELEMENT BAR INTERIOR DESIGN

Design Agency
Shanghai Taobo Decoration Design Co., Ltd.

Designer
Zhang Tao

Location
Wuxi, China

Area
300 m^2

The Element Bar is located in the new zone of Wuxi where many businessmen from Japan, Korea, America and Europe gather around. The bar offers a suitable entertainment place for these international businessmen.

Restricted by the congenital conditions, the whole area of the bar is set in a space with 3-layer elevation differences. So we choose to put the performance area in the top layer which enjoys an open view; the other two areas are in layers upper and lower, with the eye sight easily getting to the stage. The upper layer is divided into some VIP lounges while the lower one is made to be a gathering place. Every space is connected to each other while has its charm respectively. All of these are united in a magic space controlled by music.

SHELTER ISLAND PAVILION

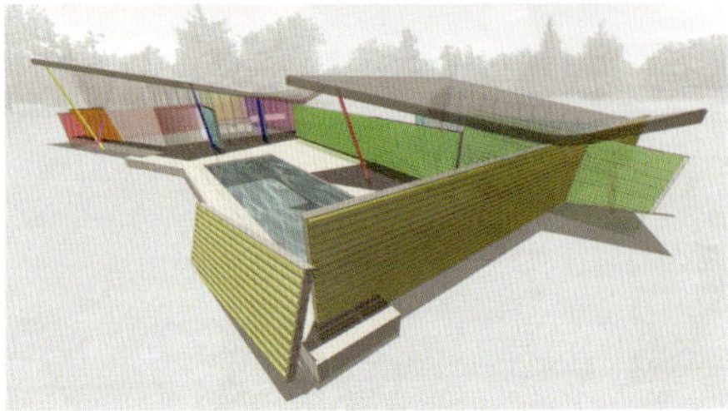

The Shelter Island Pavilion gave Peter Stamberg and Paul Aferiat of Stamberg Aferiat + Associates an opportunity to bring their influences, inspirations, aspirations and years of architectural design to bear in one place with only themselves and their budget to define the boundaries. They chose to draw on specific inspirations such as Mies van der Rohe's Barcelona Pavilion, Le Corbusier's Ronchamps, and Marcel Breuer's Wassily Chair. These works were groundbreaking and truly prescient; each was conceived in part as a prediction of the future of industrialized production and construction. Keeping the plan of the Barcelona Pavilion in mind, Stamberg Aferiat designed a house that explores the reality of the industrially-produced materials and methods of our own time. Unlike the Barcelona Pavilion that used exotic materials, they chose to utilize more common materials but rendered them striking in usage, pigment choice and detailing.

Cubists looked beyond the mechanical view of how the eye sees and employed the brain's ability to remember and anticipate, allowing one to take in a seemingly disjointed array of phenomena but still have the whole make sense. The increasing plasticity of lightweight building materials allowed Stamberg Aferiat some of the Cubists' slight-of-hand to simultaneously evoke the immediacies of built form as well as architectural dream states – the hovering roof, translucencies between inside and outside, and walls that are not walls.

Advancing material technologies have expanded the available palette through increased color intensity, optical effects and applications. Sir Isaac Newton observed the different behavior of color created with pigment and color created with light. The Impressionists and Fauves experimented with Newtonian principles to create light effects with pigment. These experiments have redefined thoughts on how colors relate to one another. Guided by Newtonian color theory, the intense palette of the house allows richly-colored reflected light to pass through translucent walls, suffusing spaces with a delighting glow.

In addition to the rigorous studies of perspective and color theory, environmentally sustainable materials and methods played a large factor in generating the design. First and foremost is the size of the project. In a time where new homes strive to maximize square footage, Stamberg Aferiat consciously kept the enclosed footprint to two small pavilions totaling 102 m^2. The home is designed for all seasons with the function of space and the areas conditioned modulated based on seasonal weather. Its heaviest use is during the summer. Large sliding doors allow indoor functions to flow onto outdoor terraces and gardens during the summer when additional space is desired and conditioned space rarely needed. The opposite occurs in winter where living occurs in a much smaller conditioned footprint. The house is one of the first on Shelter Island to use geothermal heating and cooling. Even so, it is rarely used in the warm season as Stamberg Aferiat incorporated many passive design elements into the architecture. Solid walls on the south and west side of the building block the intense summer sun while floor to ceiling translucent double polycarbonate walls allow north and east light into the space while considerably more insulation value as compared to traditional glass. Large sliding doors and windows are carefully placed to take advantage of the sea breeze to cool the interiors. Large roof overhangs provide needed shade for the pavilion interiors while defining sheltered space ideal for outdoor living.

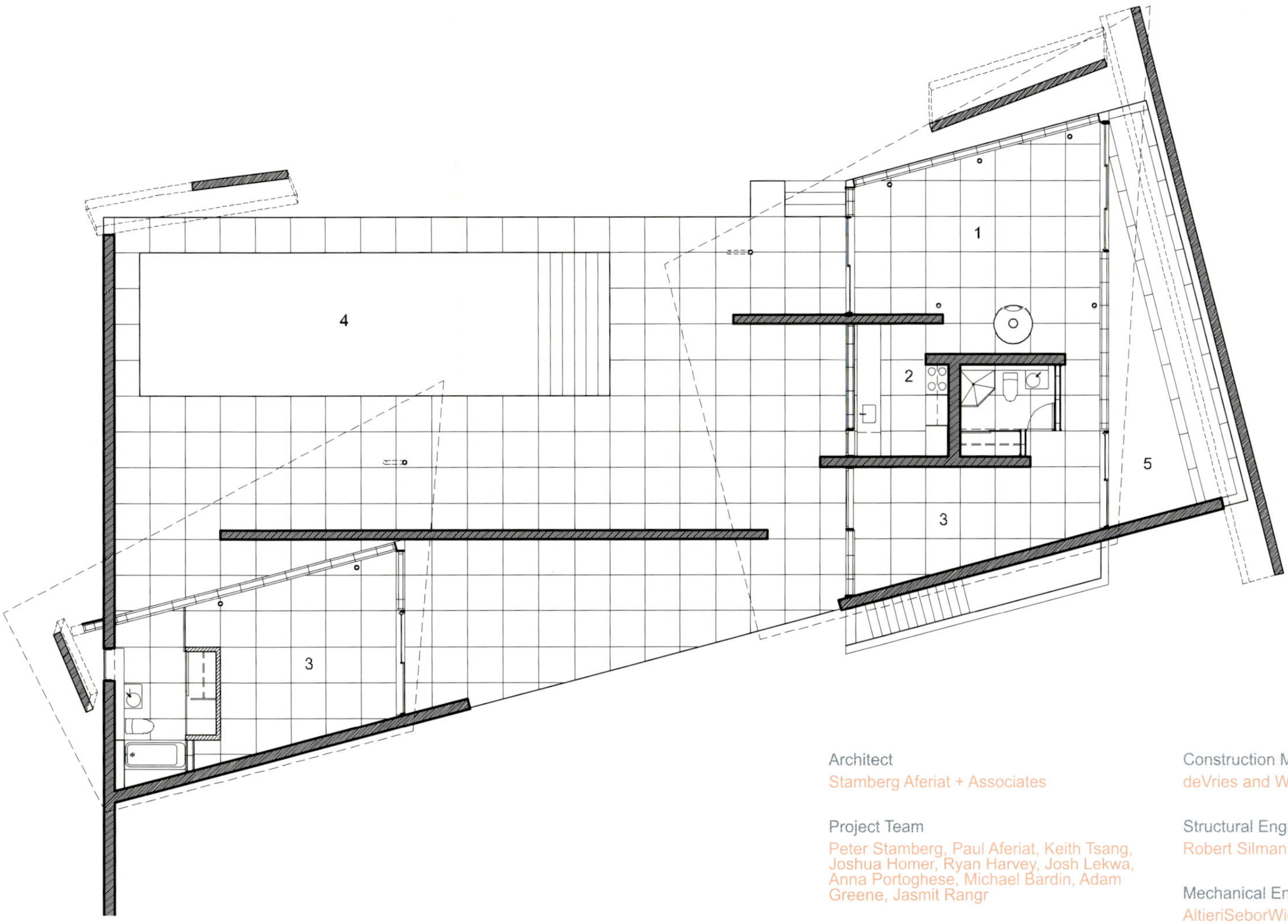

Architect
Stamberg Aferiat + Associates

Project Team
Peter Stamberg, Paul Aferiat, Keith Tsang, Joshua Homer, Ryan Harvey, Josh Lekwa, Anna Portoghese, Michael Bardin, Adam Greene, Jasmit Rangr

Location
Shelter Island, New York, USA

Area
102 m^2

Construction Manager
deVries and Wallace

Structural Engineer
Robert Silman Associates

Mechanical Engineer
AltieriSeborWieber LLC

Photographer
Paul Warchol

VENNESLA LIBRARY AND CULTURE HOUSE

The new library in Vennesla comprises a library, a café, meeting places and administrative areas, and links an existing community house and learning centre together.

Supporting the idea of an inviting public space, all main public functions have been gathered into one generous space allowing the structure combined with furniture and multiple spatial interfaces to be visible in the interior and from the exterior. An integrated passage brings the city life into and through the building. Furthermore, the brief called for the new building to be open and easily accessible from the main city square, knitting together the existing urban fabric. This was achieved using a large glass facade and urban loggia providing a protected outdoor seating area.

In this project, the designers developed a rib concept to create useable hybrid structures that combine a timber construction with all technical devices and the interior.

The whole library consists of 27 ribs made of prefabricated glue-laminated timber elements and CNC-cut plywood boards. These ribs inform the geometry of the roof, as well as the undulating orientation of the generous open space, with personal study zones nestled along the perimeter.

Each rib consists of a glue laminated timber beam and column, acoustic absorbents which contain the air conditioning ducts, bent glass panes that serve as lighting covers and signs, and integrated reading niches and shelves.

The gradually shifting shapes of the ribs are generated through adapting to the two adjacent buildings and also through spatial quality and functional demands for the different compartments of the library. Each end facade has been shaped according to the specific requirements of the site. At the main entrance, the rib forms the loggia which spans the width of the entire square. Against south/west the building traces the natural site lines, and the building folds down towards the street according to the interior plan and height requirements. On this side, the facade is fitted with fixed vertical sunshading. This shading also gathers the building into one volume, which clearly appears between the two neighbouring buildings.

A main intention has also been to reduce the energy need for all three buildings through the infill concept and the use of high standard energy saving solutions in all new parts. The library is a "low-energy" building, defined as class "A" in the Norwegian energy-use definition system. The designers aimed to maximize the use of wood in the building. In total, over 450 m^3 of gluelam wood have been used for the construction alone. All ribs, inner and outer walls, elevator shaft, slabs, and partially roof, are made in gluelam wood. All gluelam is exposed on one or both sides.

A symbiosis of structure, technical infrastructure, furniture and interior in one architectonic element creates a strong spatial identity that meets the client's original intent to mark the city's cultural centre.

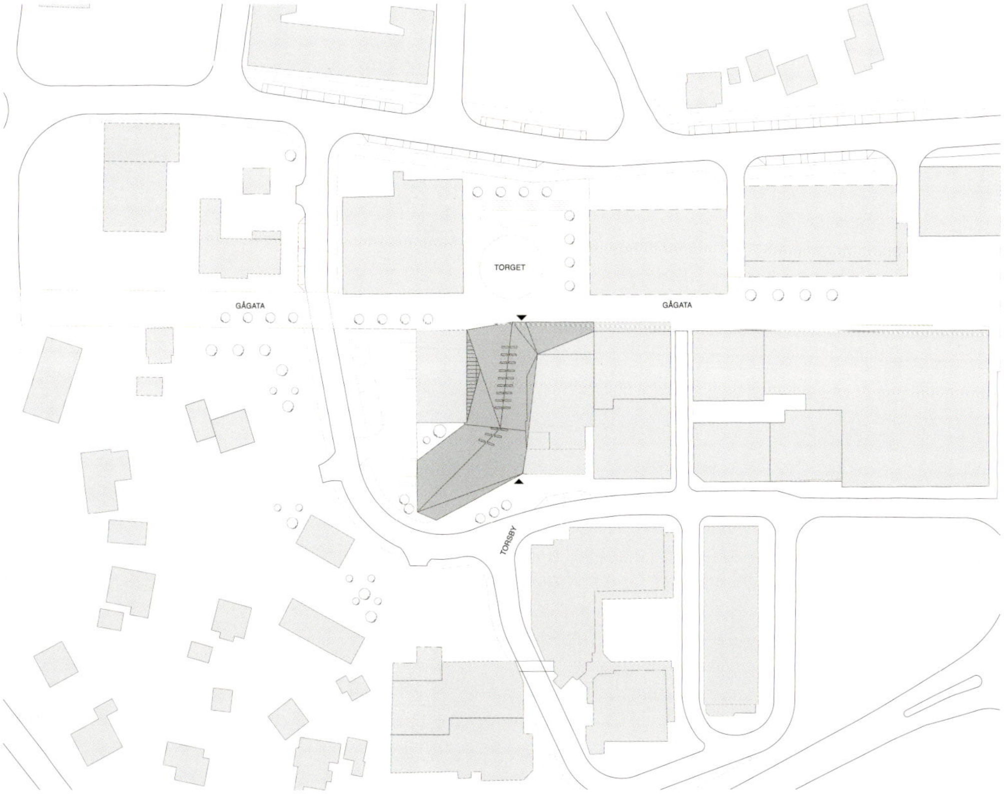

Architect
Helen & Hard

Team
Reinhard Kropf, Siv Helene Stangeland, Håkon Minnesjord Solheim, Caleb Reed, Randi Augenstein

Client
Vennesla Kommune

Location
Vennesla, Norway

Area
1,938 m^2

CALAMAR BRAND RELAUNCH & SHOWROOM

Calamar, the best-performing brand for sportswear in the Bültel International Fashion Group, demonstrates how a brand story can develop congruently in online and offline worlds with its new brand presence conceived by dan pearlman. dan pearlman's brand strategists, working together with Calamar's brand specialists, developed the new, fully-integrated brand presence over a half year period. Of particular importance here was the request to create a uniform and consistent look for all brand levels. Hence, in addition to brand strategists, communications specialists and retail designers also played a role.

Calamar seeks to maintain an active presence around the globe. Making the globe a symbol of the "new" Calamar brand is therefore a natural choice. Like Calamar itself, the globe stands for cosmopolitanism, modernity and internationality. Calamar is a little bit like a good friend who is there through thick and thin and accompanies you all over the globe – familiar, smart and, naturally, always causally dressed.

What do seasoned men have in common? They enjoy life to the fullest, and they are accomplished and yearn for more. "Staying curious" is their life motto. Men of this calibre have seen the world and their next adventure is always already planned. On their travels they wear clothing that is modern and smart, but also always functional. Such men are fans of Calamar, the urban brand for men with character.

As part of repositioning the Calamar brand, dan pearlman's communications designers kept the focus for six months on precisely this kind of men. The communications designers were asked in particular to give the brand a new visual identity. From reworking the logo, CI and the brand's online presence, to designing new lookbooks, hang tags, buttons and patches for future collection pieces no communicational medium was overlooked for the re-design.

Starting in spring 2013 the new look concept for the Calamar brand catalogue

was waiting to be discovered in stores. More than just a lookbook for the current collection, the catalogues function as travel guides, each featuring a different destination city and offering men practical advice on what goes in the suitcase for the trip.

New and smart – with the relaunch of its brand for its 30th birthday, Calamar becomes a bit more grown up while simultaneously keeping hold of the brand's proven virtues. To the delight of long-time and new customers alike, and for the success of existing and future Calamar outlets, smart prices and service will continue to be the rule in the years ahead. A blend of rational and emotional values, Calamar's smart brand core stands for the promise Calamar makes to its customers today and in the future. Here smart not only means dapper, stylish and well-dressed, but also savvy, clever and radiating confidence. From the updated collection and newly designed catalogue concept, to the well-known, sales-generating POS special offers, Calamar will continue to score additional "smartness" points in the future with its dan pearlman-designed shops.

Design Agency
dan pearlman Markenarchitektur GmbH

Client
Calamar, Bültel Group

Location
Düsseldorf, Germany

Photographer
Guido Leifhelm

CALAMAR
hattric
...MEINE HOSE

hattric
...MEINE HOSE

MARC O'POLO STORE CONCEPT 2012 – FLAGSHIPSTORE IN MUNICH

Design Agency
dan pearlman Markenarchitektur GmbH

Designer
Volker Katschinski

Client
Marc O'Polo

Location
Munich, Germany

Area
800 m^2

Photographer
Marc O'Polo

THE NEW DENIM

Marc O'Polo

THE FREEDOM
TO BE
YOURSELF

The Freedom to Feel at Home, dan pearlman creates the new Marc O'Polo flagship store in Munich.

Innovation for a new generation: the new concept for the Marc O'Polo flagship store opens the doors to a new sense of retail well-being. After more than a decade, time had come for a completely new redesign. The result is now on view in the very heart of Munich—between Theatinerstraße and the Fünf Höfe. Inside the new world of Marc O'Polo, the layout of the rooms, the interior design and lighting are choreographed to communicate "Nordic lifestyle" and highlight the philosophy of the company with Swedish roots: urban, modern and casual.

The feeling of visiting a modernized, historic apartment building soon takes over after entering the store through its two-storey entranceway. Once immersed in this world, all 800 m^2 instantly communicates a love for detail. Scandinavian designer furniture, high ceilings, carefully combined items like suitcases, photos, books and large, unexpected objects all bring the store to life: the lifestyle of the urban nomad.

The lighting concept merits special attention: reflecting off the black Nero Assoluto floor are 16,000 LED lights forming the atrium ceiling, which modulates its intensity according to daytime brightness. Inside the changing rooms, customers are bathed in a golden light and over 300 LED lights share space with a six-meter-tall waterfall of Tobias Grau lamps. Here nothing is left to chance, and no later than entering the men's world through the dressing room are the borders blurred between the retail world and the world of experience.

Sustainability is the priority and company mission of Marc O'Polo and dan pearlman, and is also evident in this project's careful use of long-lasting, energy efficient and high quality, natural materials. Here specialized retail concepts are coupled with environmental responsibility.

MATSUMOTO RESTAURANT (BEIJING)

Matsumoto Restaurant is a Japanese food chain store brand in China. This time, they invited the famous Taiwan designer Lee Hsuheng to conduct the design. Feeling people's desperate pursuit of fame and benefits in large cities like Beijing and Shanghai, the designer realizes the great importance to bring people back to simplicity and the true themselves, so the interesting concept — "pray for blessings" comes out.

Lee Hsuheng uses Japanese Taiko, Clifford board, family totem and sumo-culture elements to decorate the Restaurant. Inside, the lights create a sense of magnificence and fashion. Outside, the exterior wall, decorated by multi-layer timber Japanese Clifford boards, does not only provide a fantastic visual experience, but also presents to customers a unique space in the Restaurant to pray for blessings. The customers can write their wishes to future in the boards, which both brings the customers an expectation of realizing their wishes, and it also becomes an engagement between Matsumoto Restaurant and customers.

The designer embodies the miniature of China's harmonious society with the project. The unique design shall also produce more established customers to the Restaurant, which well expresses the designer's belief — good design shall show its own commercial value.

Design Agency
Golucci International Design

Design Team
Lee Hsuheng, Zhao Shuang, Ji Wen

Client
Matsumoto Restaurant

Location
Beijing, China

Built Area
600 m^2

Photographer
Sun Xiangyu

Design Agency
NAU | Architecture & Design

Client
Leuchtenburg Foundation

Location
Seitenroda, Germany

Photographer
Marion Lammersen

WANLI EXPEDITION EXHIBITION

The scenographic concept recasts the historic exhibition rooms as a series of interlocking air bubbles. Seven spherical and elliptical spaces constructed from stretched fabric, detail the recovery of 700,000 Ming Dynasty porcelain artifacts recently discovered off the Indonesian coast. The exhibition is built as a sensual narrative with visitors greeted by the sound of the waves at the ocean's surface, followed by gasp of a diver's breathing apparatus and the ping of sonar, as the process of maritime archaeology is explained.

Visitors are invited to become active explorers of this underwater world. Mixed in with artifacts and the personal stories of the crew are several interactive stations. "Gearing Up" offers sets of masks, fins, weight belts and diving vests that visitors can try on. In a darken film room, two mounted "Search Lights" are fitted with ultraviolet spotlights; when guests graze the oval walls with UV light, facts and figures about the dive operations suddenly become visible, inviting playful interaction.

The exhibition engages through its use of media in the form of narrated film segments, rooms washed with atmospheric projections, and a digital logbook. As the excavation in Indonesia proceeds, information about the dives and weather conditions from the ship's daily logs are remotely updated and displayed in the exhibition. This makes for an experience which is highly contemporary and always offers visitors something new.

AUF DER
SUCHE
THE SEARCH
ARQUEONAUTAS
WORLDWIDE

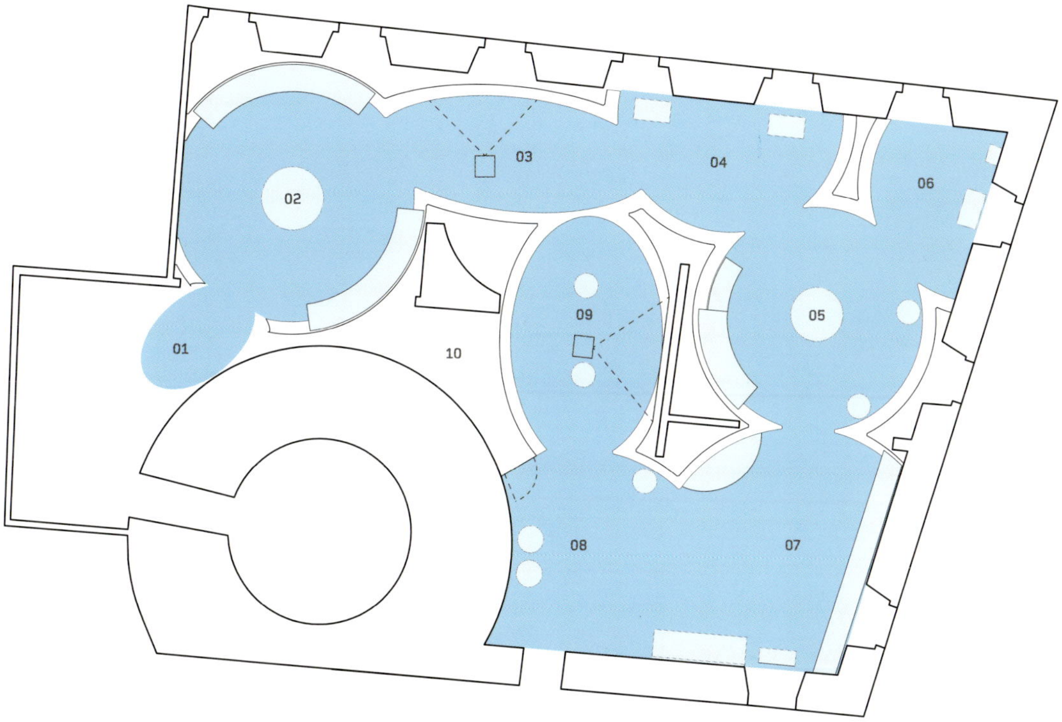
01
02
03
04
05
06
07
08
09
10

WANLI-PORZELLAN
WEISSES GOLD
VOM
MEERESGRUND

ELTHANDEL
M 1600
LOBAL TRADE IN 1600

24.000
1622.

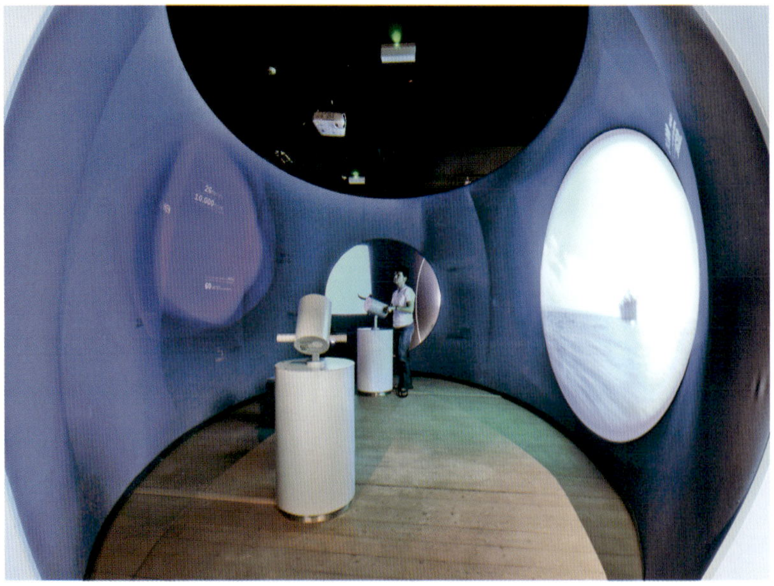

3,80
Wanli Fundhöhe
Wanli Porcelain Mound

ARQUEONAUTAS
DAS UNTERNEHMEN

FUND
3,80 m

OCTOPUS TENTACLES

Interior design of three-room apartment was made for a family of customers (parents and a primary school age girl). The volume of open-space combines lounge and dining/cooking area representing a day zone and main accent.

The digital component of the project has its roots in meta-balls research. Places that need artificial lighting most are taken as control points, each with its own value, proportional to significance and connected with each other, creating a smooth flow. There is a flow between door panels in the hall connected with the dynamic floor lines,continuing the trend of the ceiling.

The parametric rhodoneacurves research is taken as a basis in furniture details. Curves act as a differentiator: the number of petals depends on the roomsize, the smaller the room is the less petals it gets.

The project is experimental site for using 5-d machine and properties of artificial stone used for a bar table and curved construction for wine storage.These two forms flow smoothly into each other demonstrating the benefits of advanced materials and digital modeling / production.

Design Agency
Dmytro Aranchii Architects

Project Team
Dmytro Aranchii,
Mariia Aranchii

Client
Private

Location
Kyiv, Ukraine

Area
115 m^2

WAKUWAKU DAMMTOR

WakuWaku has relaunched itself as a fast food restaurant and organic food store in one. The brand values "organic" and "sustainable" remain the clear focus of all communication.

The large, open façade provides an unrestricted view of a space, which is almost entirely encased in solid wood panelling, creating the ideal stage to display the products in the WakuWaku world. One side wall with floor-to-ceiling shelving integrates both display compartments and glass-fronted refrigerators. A long central counter and parallel niches provide seating spots for all communicative requirements. Different shapes of chair break up the seating landscape and hark back to the original WakuWaku outlet — as do the chair legs, dipped in the WakuWaku corporate colour. The untreated wood dominating the space is synonymous with the chain's ecological sustainability. The rough wooden surfaces contrast with the intricate wall sketches created by Chris Rehberger using taut strings and the canopy of fine wire lamps.

Design Agency
Ippolito Fleitz Group – Identity Architects

Project Team
Peter Ippolito, Gunter Fleitz, Moritz Köhler, Michael Bertram, Markus Schmidt, Timo Flott

Location
Dammtor str. 29-33, Hamburg, Germany

Area
145 m^2

Photographer
WakuWaku/Benjamin Nadjib

BIO

GOOD FOR EARTH
LEMONAID

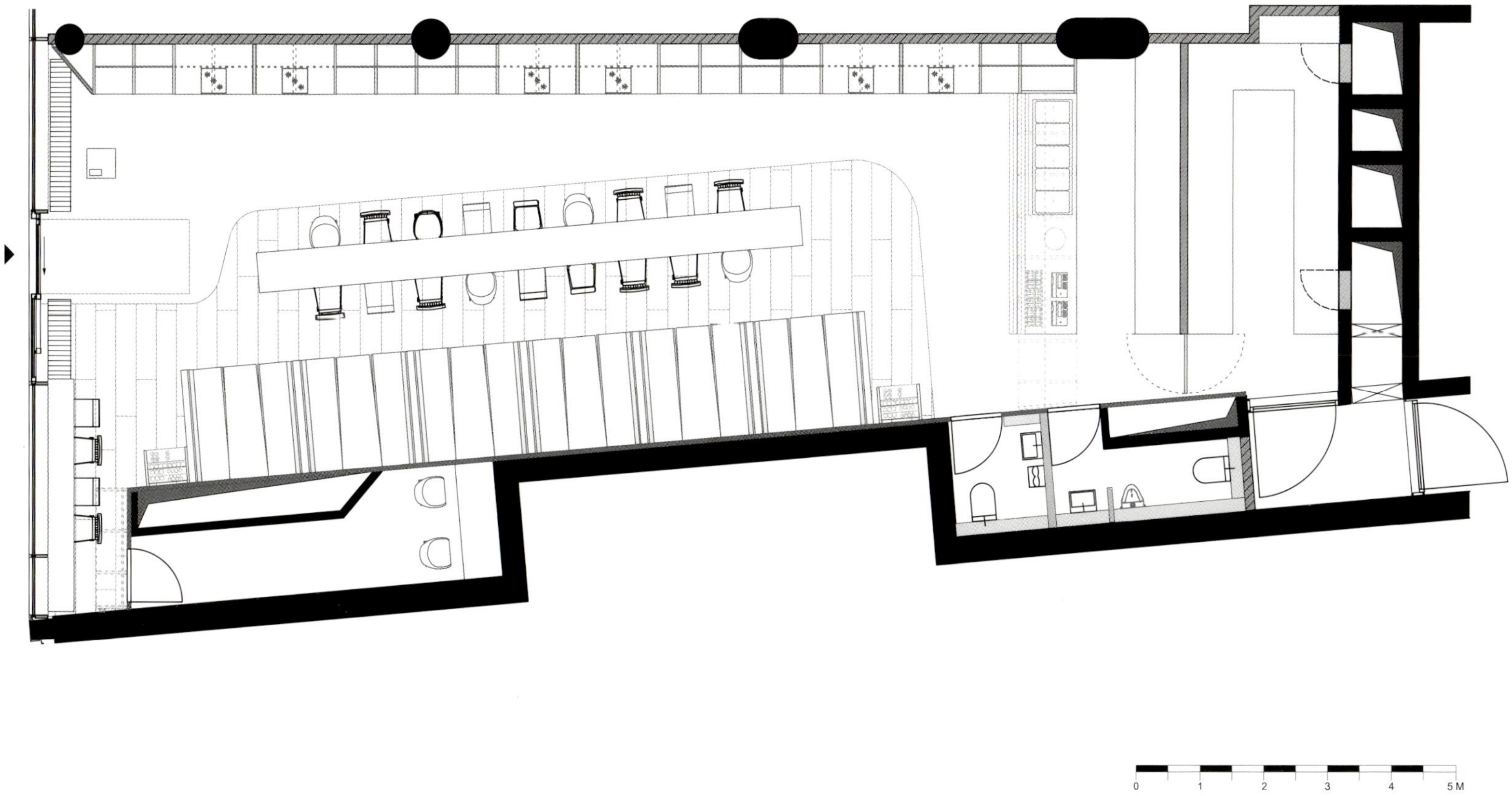
0
1
2
3
4
5 M

Architect
Manuelle Gautrand

Client
Automobiles Citroen

Location
Paris, France

Photographer
Philippe Ruault, Jimmy Cohrssen

CITROEN FLAGSHIP SHOWROOM

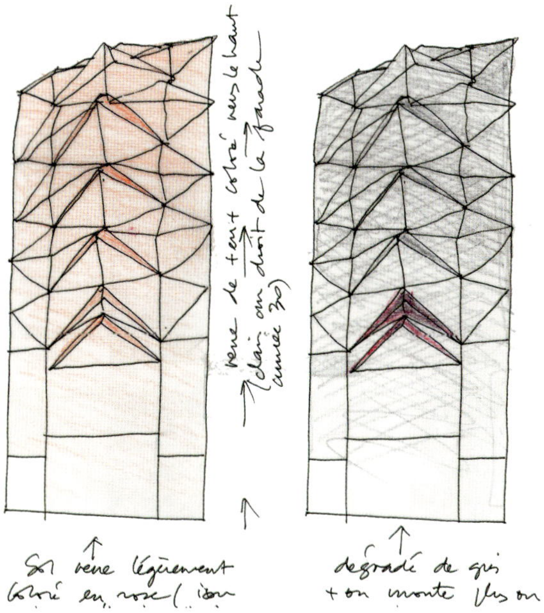

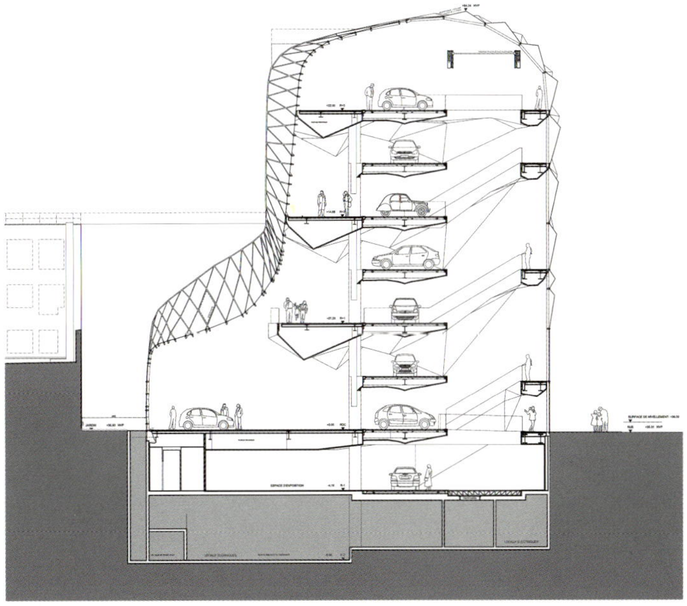

The new Citroën showroom is at number 42 Champs Elysées; Citroën have owned the site since André Citroën set up shop there in the 1920s. His original showroom was beautiful, the interior was extremely theatrical, and the glass rectangle façade beautifully proportioned, very minimalist and contemporary.

On street level, the glass façade is minimalist and demonstrates a certain rigour with its flatness and use of large rectangles, but the introduction of the chevron signals the start of some much more original design, with lozenge shapes, triangles and chevrons. The higher up the building looks, the more three-dimensional it becomes with the introduction of prisms that bring new depths to the design. Finally, the top section of the new building is like a great glass sculpture, recalling origami in its complexity. The chevron remains present yet discreet, becoming less defined and more suggested in the overall form, and almost subliminal, in this exciting project, midway between a building and a fine art sculpture.

The designers originally conceived the use of red, the brand's signature colour, in the glass panels but they decided it would be too bright from the outside. There were some concerns about the building not harmonising with its neighbours on the Champs Elysées, so they've created a filter that on first sight, masks the red colour from the exterior. This totally original filter, which is cleverly constructed inside the finished glass, also minimises the heat of the sun passing through, and will also create a diaphanous pearly white atmosphere inside the building. The red colour can still be seen from the inside of the building, reflecting the brand's signature colours.

The main role of the building is as a place to show cars, and designers wanted to express this primary aim in the form of the space itself. The shape of the building itself is inspired by the shape of a car, it's not an object with a front, a roof or a rear, but something moulded with curves and fluidity, that links the front, roof and rear with a continuity that is like the form of a car itself, creating unity between the place and the product, and makes a rich and complex interior.

To display the cars themselves, attached to a central mast are eight circular platforms each of which takes a car. The platforms are 6 m in diameter, and each one turns to show off the car on all sides and has a mirrored base to reflect the car below. Around the display, the public is led by a series of staircases and walkways that spiral past the cars. The designers were trying to create something like a museum or a cultural building, a space which would encourage people to spend time there. There is a panoramic lift to take people to the top of the building, and they are able to enjoy an exceptional view of both Paris and the sky.

CITROËN

CITRO
MÉTISSE
CITROËN

TROLL WALL VISITOR CENTRE

Architect
Reiulf Ramstad Arkitekter

Client
Private

Location
Trollveggen, Møre og Romsdal, Norway

Area
700 m^2

Photographer
Dag Terje Alnes, Reiulf Ramstad Arkitekter

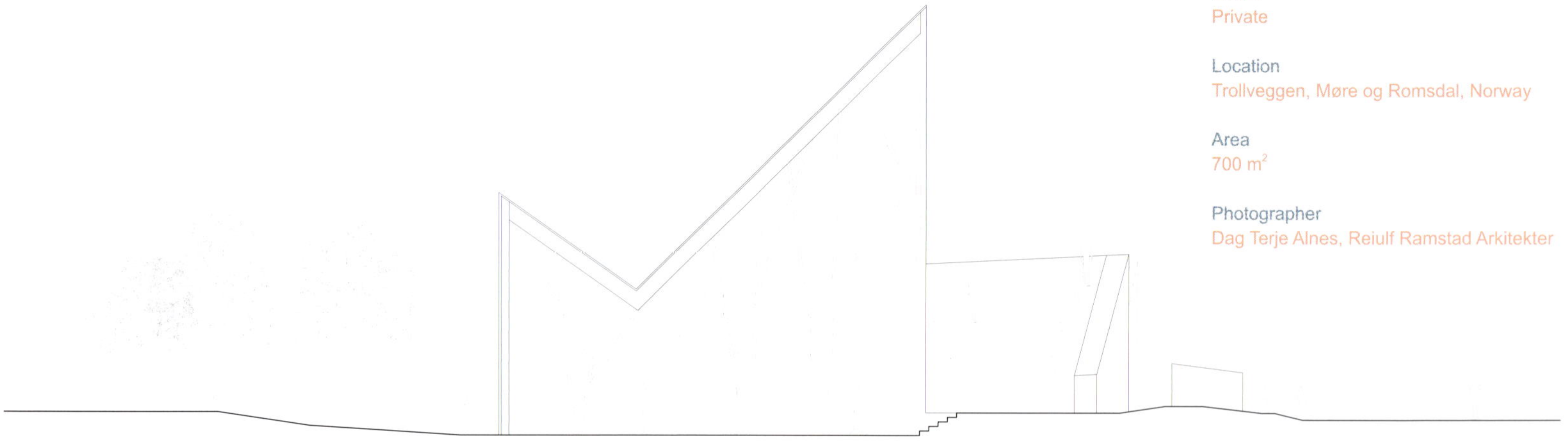

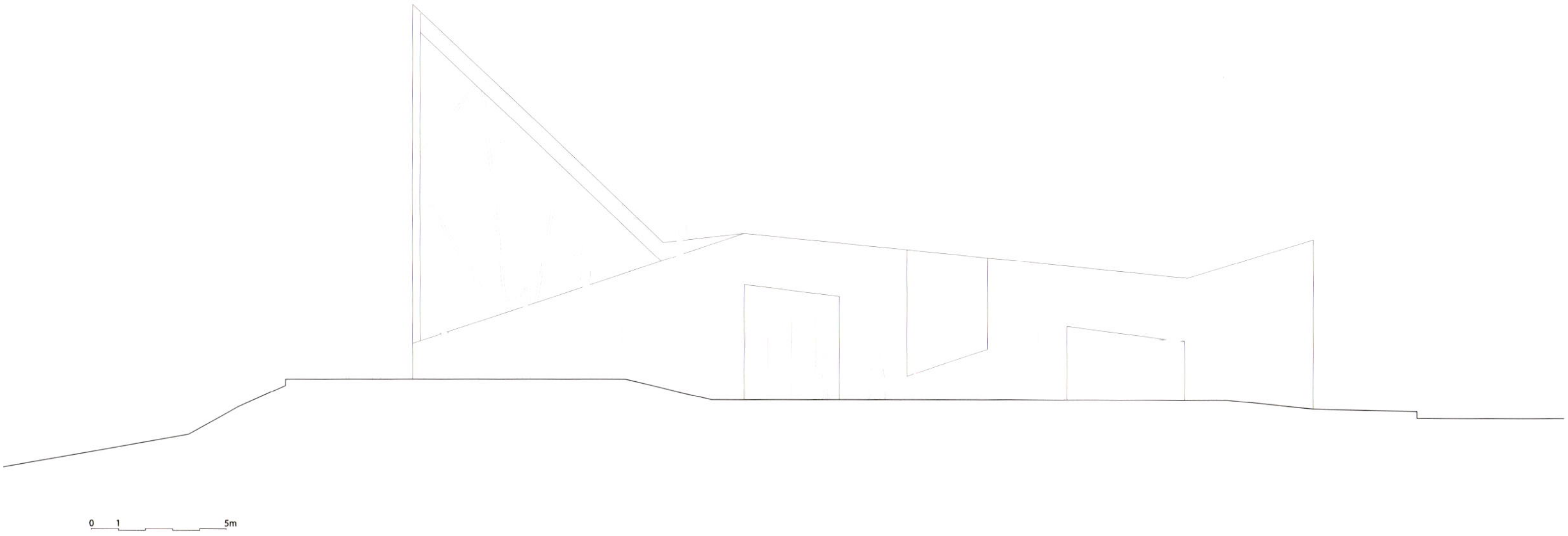

It's a new cursor at the foot of the Troll Wall. The architecture of the new visitors centre next to E139 is an outcome of the site's close relation to the impressive mountain wall, one of Norway's many nature attractions. The building has a simple, though flexible plan, with a characteristic roof that has its character from the majestic surrounding landscape. These simple ways of design gives the building its character and identity that makes the service centre an eye-catcher and an architectural attraction in the region.

Trollveggen
besøkssenter
Kafeteria / Trollveggen utsikt / Trollveggen film / Souvernir

Contributors

4N Architects Ltd.

4N was founded by Sinner Sin and Danny Ng, who describe themselves as "hands-on architectural designers". They graduated from the University of Melbourne and have design experience across Asia Pacific to Middle East, including Hong Kong, Macao, Mainland China, Singapore, Dubai and Australia. 4N's strengths are building and spatial design, sustainability design and total design. At the Preliminary design stage, we work intensively with clients in brainstorming and developing concepts. We aim to establish trust with the client – something which is a major priority for us. The original idea was to use design as our means of expression for giving back to the community. Architectural and interior design have been a medium for merging art and building technology without conflict. As a city develops, the principle tends to become neglected; we have a vision of reintroducing it back to the public.

AIDIA STUDIO

AIDIA STUDIO is a platform of experimental design in architecture, urbanism, photography and other creative disciplines.

The work spans across scales and typologies, from the domestic to the urban striving always for specific responses, performing at different levels and with intricate and sophisticated beauty.

AIDIA STUDIO was founded by Rolando Rodriguez-Leal and Natalia Wrzask in 2011 as a sandbox for the work carried out in different arenas. It has a presence each in Beijing, London and Mexico City.

Ana Milena Hernández Palacios

Ana Milena Hernández Palacios was born in the Colombian city Bogota the 28th of August, 1982. From young age on she developed a creative talent participating in the decorative events from her mother. It was in 2001, at the age of 18 that she decided to start an adventure in Europe moving to Spain. There she worked the first years as a window dresser and responsible of merchandising for the stores of Maximoda. However as she was fascinated by the world of interior design she decided to start studying interior design at the well known designschool of Valencia Barreira. After finishing her studies she worked for a while in an architecture studio where she realized very soon that she wanted to develop her own style. The entrepreneurial spirit of her parents always had been very present and that's what convinced her without a doubt to launch +Quespacio. The aim with +Quespacio is very defined, design creative spaces taking in count the needs from her clients and involucrate the diferent areas of design with the goal to create added value for their establisments.

AY Architects

AY Architects is founded on the cross-fertilisation of architectural practice, teaching and research. It is an open form of practice involving a variety of processes and the interaction of clients, consultants, artists and trades.

They are interested in how space is experienced, how it relates to time, and what it actually means to people. Their projects are dissimilar, ranging from small public, educational and private buildings, ideas' competitions about urban space, and exhibitions. This diversity enables us to test ideas in different contexts and build a layered open-minded attitude to design. On this basis them approach each project afresh and place an emphasis on what makes it unique.

AY Architects was formally set up in 2005, although its founding partners, Yeoryia Manolopoulou and Anthony Boulanger, first started working together in 1998 as postgraduate students at the Bartlett School of Architecture, London. They went on to co-found the practice Tessera that ran from 1999 until 2003. Anthony also worked with Ian Ritchie Architects in London from 1998 until 2005. He is a Lecturer at the University of Westminster, where he has taught Diploma Studio 16 since 2007. Yeoryia is a Senior Lecturer and Director of Architectural Research at the Bartlett. She is also a PhD supervisor and teaches MArch Unit 17.

Ball-Nogues Studio

Ball-Nogues Studio is an integrated design and fabrication practice operating in the territory between architecture, art, and industrial design. Essential to each project is the "design" of the production process itself. They devise proprietary systems of construction, create new fabrication devices, develop custom digital tools, and invent materials with the aim of expanding the potential of the physically constructed world. They share an enthusiasm for the fabrication process as it relates to the built world both physically and poetically by letting the properties, limitations, and economic scenarios associated with a process guide a structure's ultimate form while developing methods to extend the intertwined boundaries of aesthetics, physical performance and lifecycle.

Speculation and execution are inexorably linked in their work; each project demands that they maintain tight control over design and production. As young practitioners, this requires a do-it-yourself ethos. Consequently, they have "designed" our career so they can exploit opportunities to build that are outside the constraints of the conventional architectural milieu. Although their projects are experimental with respect to production, they are far more than prototypes; each directly addresses human occupation by enhancing and celebrating social interaction through sensation, spectacle and physical engagement.

Carlos Arroyo Architects

Carlos Arroyo Architects is a Madrid based architecture & planning office of international scope, currently building in Spain, France, Belgium, Argentina and Rwanda.

Their multidisciplinary team conducts projects of a varied nature, with a strong emphasis on innovation and sustainable development; innovation on all scales, from building technology to landscape management, developing new types of public building, or researching into new forms of housing.

Sustainability is an essential driving force of our work, which has been described by critics as "sustainable exuberance".

Charlotte Mann

Charlotte Mann is a British artist who makes 1:1 scale black line drawings of objectively perceived interior and exterior spaces. These drawings often take the form of murals others are room sized installations.

In spite of the severely restricted media and many rules governing the choice and treatment of subject matter her work has a expansive generous feel to it, she says of her process: "The deliberate absence of color from the form of my work correlates not to an indifference to or dismissal of color but to the importance played by the subject of color, conceptually, this is also how I feel about three dimensional form and time. I am aiming through my work to engage with the physical presence of a conscious being in an environment, the most essential element, the perceiving person is never drawn. That presence can only be provided by the viewer. That is the basic rule around which my practice is formed, and other decisions follow from it."

Cristina Parreño Architecture

Cristina Parreño Alonso is a licensed architect in Spain and UK, with ten years of professional experience working in Madrid and London. She got a Masters and Bachelor degree in Architecture and Civil Engineering from the University of Madrid, EscuelaTécnica Superior de Arquitectura (ETSAM) where she graduated with honors.

She started her professional career in Madrid. Later she moved to London and joined Foreign Office Architects (FOA), where, as project director she was responsible for a number of projects in Spain and UK for several years.

Since 2009, she has been working on her own practice and teaching at various architecture universities. She has taught design studio at the University of Western Australia, at the State University of New York at Buffalo and she currently teaches graduate and undergraduate design studios at MIT School of Architecture and Planning.

She has won several architectural competitions, among others, the 1st prize for the Urban Rehabilitation of the Business complex AZCA in Madrid, award for the Housing Competition for Young Architects J5 in Andalusia and Honorable Mention for the National Library of Slovenia. Her work has been published in newspapers like El País and El Mundo and in several architectural magazines like Quaderns, Via Arquitectura and Arquitectura COAM.

dan pearlman

dan pearlman is a strategic creative agency in the areas of brand strategy, brand communications, brand architecture and experience architecture.

As an interdisciplinary, holistic agency, dan pearlman combines strengths in different areas: Branding, Research and Innovation, Internal Branding, Brand Experience, Fairs and Events, Retail, Visual Communication, 3D and Motion, Public Relations, Hospitality, Leisure, and Zoo.

Since the establishment in 1999, dan pearlman has been trusted by many national and international clients, who cooperate with the agency. Among the main customers of the Berlin-based agency with 60 employees are: Allianz, BMW, Lufthansa German Airlines, Mercedes-Benz, MINI, Roca, Qatar Railways, Rene Lezard, Karstadt and Marc O'Polo.

dan pearlman was honored several times for many various projects. The honors of particular note are: the Design Award of the Federal Republic of Germany 2011 in gold for the Lufthansa Brand Academy and the Store of the Year 2011 Award for the design and development of brand and store concept for COEO.

Dmytro Aranchii Architects

DMYTRO
ARANCHII
ARCHITECTS

Dmytro Aranchii founded Dmytro Aranchii Architects in 2007 as the digital design research pioneer practice in Ukraine.Studio works in fields of algorithmic architecture, parametric and generative design.

The design agency prefers a method that replaces the style in design and architecture. Combining artistic researches and algorithms of computer programming architects realize an immense potential of theoretic component closely related to practical activity. That is why a great attention is paid to algorithmic methods of morphogenesis that allow getting reasonable results.

Nowadays Dmytro Aranchii Architects is a team of young progressively orientated specialists that believe in future of parametric architecture.

Dopludó Collective

Dopludó Collective was founded in St. Petersburg, now based internationally, design studio founded in 2005. They do a wide range of visual disciplines, such as design, murals, independent art & public projects. The name "Dopludo" apparently comes from a phonetic spelling of the French "deux plus deux" (two plus two). They take that to be a kind of shorthand for the whimsical notion that 2+2=5, an idea further abstracted by the fact that there are only three people in the collective. We are inspired by such a phenomenas, as deliberate conscious infringement of logic rules. Two plus two might equal not only four or five, it could be decided on your own, and starting from this point moved further, how far? – Everyone decides himself.

EDGE

EmmiKeskisarja and PekkaTynkkynen (www.pekkatynkkynen.com) are independent architects working through EDGE Laboratory for Urban and Architectural Research of the Tampere University of Technology School of Architecture. EDGE serves as research infrastructure provider, supports when looking for funding channels and assists in the preparation, administration and management of research projects at starting and on-going levels.

Francisco Sarria / Studio

Francisco Sarria / Studio is a design practice based in Copenhagen that works in an innovative and multidisciplinary manner combining different design disciplines and skills to approach a variety of design projects: from furniture and interiors to branding, exhibition and retail design.

The studio takes brand concepts to create tangible, physical spaces. The studio's work encompasses furniture, retail, workspace, exhibition, brand experience, pop-up, restaurant and bar design. The studio finds solutions for the clients in many different ways without following a set formula as each project has it's own needs.

Francisco Sarria / Studio produces groundbreaking spaces, objects and graphics with a strong geometric and sculptural aesthetic for businesses and private customers alike.

Frank Tjepkema

Tjep. (Frank Tjepkema, 1970) grew up in Geneva, Brussels and New-York. After experiencing a variety of cultures during his youth he settled in the Netherlands in 1989 to study industrial design. He graduated Cum Laude from the Design Academy in Eindhoven in 1996.

Tjep. set up his own design studio called Tjep. together with Janneke Hooymans in 2001. Tjep.'s field of work has developed to include products, furniture, accessories, interior design and interior architecture for a multitude of restaurants. Tjep' s work can be found in the world's most influential galleries such as Moss in New-York (Bling Bling, Heartbreak...) all the way to the shelves of your refrigerator (Waater bottle).

FreelandBuck

FreelandBuck is an architectural design practice based in New York and Los Angeles affiliated with Yale and Woodbury Universities. Our office focuses on research and design, exploring the overlap between academia and practice.

David Freeland is the principal of FreelandBuck in Los Angeles and adjunct faculty at Woodbury University. With over 10 years of experience in architecture, he has worked on award winning projects with a number of offices in New York and Los Angeles including Michael Maltzan Architecture, Roger Sherman Architecture and Urban Design, RES4, AGPS, and Eisenman Architects. He is a graduate of University of Virginia and the UCLA Department of Architecture and Urban Design.

Brennan is the principal of FreelandBuck in New York and a Critic at the Yale School of Architecture. From 2004 to 2008 he was assistant professor at the University of Applied Arts, Vienna teaching with Greg Lynn. He has practiced both landscape architecture and architecture, having worked for Neil M. Denari Architects and Johnston Marklee & Associates in Los Angeles. He is a graduate of Cornell University and the UCLA Department of Architecture and Urban Design.

Garth Britzman

Garth Britzman, of Brookings, South Dakota, is a senior Architecture major at the University of Nebraska-Lincoln. Britzman is interested in sustainability as it relates to urban and architectural design. As a student, Britzman is involved in various committees and organizations that promote sustainability on the university campus as well as educate students about sustainable lifestyles. Britzman is an entrepreneur working to grow businesses in the fields of industrial design and sustainability consulting. He believes that creativity in design inspires innovation.

Gerry Judah

Gerry Judah was born in 1951 in Calcutta, India and grew up in West Bengal before his family moved to London when he was ten years old.

He studied Foundation Art and Design at Barnet College of Art (1970 – 1972) before obtaining a First-Class Honor degree in Fine Art at Goldsmiths College, University of London (1972 – 1975) and studying sculpture as a postgraduate at the Slade School of Fine Art, University College London (1975 – 1977). He went on to build a reputation for innovative design, working in film, television, theaters, museums and public spaces, creating settings for productions at many international museums.

Gemelli Design Studio

Gemelli Design is a creative art and design Sofia-based studio with driving force the experienced interior designers Branimira Ivanova and Desislava Ivanova. The main aim of the designers is creating of unique residential and commercial interiors. They achieve uniqueness through transformations of the spaces with the use of proportions, forms and colors.

Desislava Ivanova and Branimira Ivanova were born in Sofia, Bulgaria in 1979.

They received their Bachelor degree in Interior Design from Kingston University, London, and their Master degree in Interior Design from Florence Design Academy in Florence, Italy.

Their first touch with interior design is in the School of interior architecture and woodworking in Sofia. Then it turns into their passion. In order to develop their abilities and to improve their knowledge, they follow a succession of studies in University of Forestry, Sofia; Stuttgart University of Applied Sciences; College of Furniture Design, Austria; Kingston University, Florence Design Academy.

Golucci International Design

Golucci International Design was established by Taiwanese Designer Lee Hsuheng in 2004. Our highly motivated and qualified designers fully recognize the importance of professional acumen.

Each project is conceptualized and developed by our experienced design team. Over the years, our works have included a wide range of Clubhouses, Hotels, Bars & Restaurants. Our approach to management ensures a high quality end product and we express the essence of our creative ideas to the best benefits of our clients.

heri&salli

Since 2004 work Heribert Wolfmayr (1973) and Josef Saller (1971) – heri&salli- out spatial drafts, architectural horizons; they work with interventions, bound with surroundings and landscapes which reach their targets in materiality compared with people. By opposites, put into interlinked connections, an architectural idea - as a collection of different substantial barriers and surfaces – reaches its concept and necessity in connection with material, open space and human being. Man as active being always is the reason for possibilities of architectural drafts.

HENN

HENN is an internationally operating German architectural practice with more than 30 years of building expertise in the fields of culture and administration buildings, education, research and development and production buildings.

Projects such as Die Autostadt, Die Gläserne Manufaktur and the Project House in the Research and Innovation Center of BMW have been internationally acclaimed; ongoing large-scale projects in China include the headquarters for the two biggest life insurance companies China Life and Taikang Life, a production plant for BMW in Shenyang and the Science City of Nanopolis in Suzhou.

As a general planning practice HENN provides experience in all work phases. The broad scope of work includes architectural planning, interior design, master planning, quantity surveying, construction management and LEED certification.

The office is managed by Gunter Henn and nine partners. These days, 330 architects, designers, planners and engineers work in project teams in our offices in Munich, Berlin, Beijing and Shanghai.

HENN has long years of experience in the research and development of innovative spatial concepts and of efficient design solutions. Key factor for the success of any project is the precise examination of the complex requirements of every design task.

As an internationally operating architectural office HENN command expertise in all design and building phases. The broad scope of work covers the fields of masterplanning, architectural design, interior design, quantity surveying, construction management, general planning, leed certificate and programming.

Helen & Hard

HELEN & HARD

Helen & Hard was founded in 1996 in Stavanger on the west coast of Norway by Norwegian architect Siv Helene Stangeland and Austrian architect Reinhard Kropf. Today, the company has a youthful staff of 20 drawn from 8 different countries, with offices in both Stavanger and Oslo.

We design in different scales and scopes, on a wide spectrum of projects ranging from single family houses to large public buildings, from offices and multi-family housing projects to master planning.

We aim to creatively engage with sustainability, not only in the design of spaces, but also in the conception and organization of the design process, including construction and fabrication. Our goal is move away from a solely technical and anthropocentric view, allowing the project to unfold in relation to its physical, social, cultural and economic context.

Iñaki Echeverria

Iñaki Echeverria is an architect and landscape urbanist based in Mexico City. His eponymous firm, founded in 2008, has been awarded numerous high profile commissions, both public and private such as Texcoco Lake Ecological Park (143 million square meters.).

Echeverria's multidisciplinary approach provides unique and specific solutions to complex conditions ranging from grand scale environmental remediation, public reclamation projects and landscape urbanism planning to design, architecture and art installations. In these investigations a bottom-up collaborative approach to design is always privileged.

Currently the firm has been invited to its first commission in Asia, a 500 acres landscape urbanism plan for Wuxi outside Shanghai; is working with the largest luxury retailer in Mexico and designing technological centers nationwide for a national employer's association. The practice's scope includes international competitions and proposals such as the Taichung Gateway Park, the Pfeffeberg Underground in Berlin or a public space strategy for Ciudad Juarez. The office also engages private commissions like the "Tres Vidas" House in Acapulco, Amapas I Development in Puerto Vallarta or Sonora offices in hip-neighborhood Condesa.

Institute of Building Structures and Structural Design

itke

The Institute of Building Structures and Structural Design (ITKE) at the University of Stuttgart focuses its activity on the development of structures as the main aspect of architecture. Combining teaching and research in a highly interdisciplinary environment, ITKE's goal is to push the boundaries of engineering design and material science towards new and non-standard applications in the field of architecture.

The two main research interests of the Institute are geared towards material science for the production of high performance materials and their application, along with structural morphology and the study of innovative structural systems. These fundamental aspects of the research activities at ITKE are investigated both from a theoretical and a practical point of view, integrating computational engineering and advanced analysis methods together with technological fabrication and development of full scale prototypes.

The synergy between research and teaching environment is a central feature of the work carried out at ITKE. The interaction between scientific investigation and academic activities provides a privileged platform to develop high quality research and to establish solid collaborations with industry partners, allowing the continuous trade of original ideas and an innovative approach to structural design which redefines the border between architecture and engineering.

Institute of Computational Design

Stuttgart University

Institute for Computational Design
Institut für Computerbasiertes Entwerfen

The Institute for Computational Design (ICD) is dedicated to the teaching and research of computational design and computer-aided manufacturing processes in architecture.

The ICD's goal is to prepare students for the continuing advancement of computational processes in architecture, as they merge the fields of design, engineering, planning and construction. The interrelation of such topics is explored as both a technical and intellectual venture of formal, spatial, constructional and ecological potentials. Through teaching, the ICD establishes a practical foundation in the fundamentals of parametric and algorithmic design strategies. This provides a platform for further exploration into the integrative use of computational processes in architectural design, with a particular focus on integrative methods for the generation, simulation and evaluation of comprehensive information-based and performance oriented models.

There are two primary research fields at the ICD: the theoretical and practical development of generative computational design processes, and the integral use of computer-controlled manufacturing processes with a particular focus on robotic fabrication. These topics are examined through the development of computational methods which balance the reciprocities of form, material, structure and environment, and integrate technological advancements in manufacturing for the production of performative material and building systems.

Ippolito Fleitz Group

Ippolito Fleitz Group is a multidisciplinary, internationally operating design studio based in Stuttgart. Currently, Ippolito Fleitz Group presents itself as a creative unit of 37 designers, covering a wide field of design, from strategy to architecture, interiors, products, graphics and landscape architecture, each contributing specific skills to the alternating, project-oriented team formations. Their projects have won over 160 renowned international and national awards.

J. MAYER H. Architects

Founded in 1996 in Berlin, Germany, J. MAYER H Architects' studio, focuses on works at the intersection of architecture, communication and new technology. From urban planning schemes and buildings to installation work and objects with new materials, the relationship between the human body, technology and nature form the background for a new production of space.

Jürgen Mayer H. is the founder and principal of this crossdisciplinairy studio. He studied at Stuttgart University, The Cooper Union and Princeton Universtiy. His work has been published and exhibited worldwide and is part of numerous collections including MoMA New York and SF MoMA. National and international awards include the Mies-van-der-Rohe-Award-Emerging-Architect-Special-Mention-2003, Winner Holcim Award Bronze 2005 and Winner Audi Urban Future Award 2010. Jürgen Mayer H. has taught at Princeton University, University of the Arts Berlin, Harvard University, Kunsthochschule Berlin, the Architectural Association in London, the Columbia University, New York and at the University of Toronto, Canada.

Jean de Lessard

With more than 20 years' experience, Montreal designer Jean de Lessard has designed numerous interiors, particularly in the commercial and deluxe residential sectors. His eclectic skill has led him to work on spaces with very diverse vocations, from the corporate offices of the world-class company Acquisio to the Cache-à-l'eau recreational centre for children, Hotel Trylon, and the redesign of the interiors of the restaurants in the Rouge Boeuf chain. He is regularly asked to travel to Europe, the Middle East, and Asia to analyze different interior design mandates.

In 2010, Jean de Lessard received the prestigious International Best Interior Design Americas award in London, England, for his design of the offices of the marketing agency Upperkut. He has also won the Prix Intérieurs | Ferdie, in the Commerce Design Montréal competition, and awards from the Institut Design Montréal. He is a member of the Club des Ambassadeurs des Grands Prix du Design, and his designs have been published many times around the world in gold-standard interior design books and magazines.

Kengo Kuma

Kengo Kuma, first class Architect in Japan. Born in Yokohama, Kanagawa Prefecture, Japan, 1954. He completed the Master Course at The University of Tokyo in 1979. He established Spatial Design Studio in 1987, Kengo Kuma & Associates in 1990 and Kuma & Associates Europe (Paris, France) in 2008.

Kitsch Nitsch

Kitsch Nitsch is David Kladnik and Jaka Neon. They studied graphic design at the Ljubljana Academy of Fine Arts and Design, worked briefly in advertising and started their own studio in 2006. Originally working in the field of graphic design under the name of Pop-up design studio, they started the Kitsch Nitsch project a year later in order to have a playground where they could explore the less rigid aspects of design such as decoration. Currently they devote most of their time to interior decoration, scenography and illustration. They believe in style, the importance of personal preferences in the design process and that one can find space for humor, whatever the project might be. They have an artistic approach to their work and like to mix different esthetics into a harmonious whole or a clash of contrast. They work a lot with self adhesive vinyl film that has the ability to quickly transform the visual aspect of a room or an object without changing its internal structure. They prefer strong imagery over subtle corrections, the kitsch part of their name holds no love for tackiness but rather describes the will to create personalized and unique visual messages instead of resorting to prefabricated standardized solutions.

Laboratory for Explorative Architecture & Design Ltd.

Laboratory for Explorative Architecture & Design Ltd. (LEAD) is a young Hong Kong & Antwerp based architectural design and research practice, founded by Kristof Crolla & Sebastien Delagrange. LEAD explores innovative architectural outcomes by strategically combining advanced digital design and fabrication technology with traditional manufacturing opportunities and on-site craftsmanship. Its scope and services stretch from initial design through construction up to project realisation. In its work LEAD seeks to provoke sensual, haptic reactions from users interacting with the architecture. Within its first two years of operation LEAD won the Hong Kong Global Design 2011 Excellence Award for their Shine Fashion Store, received the 2012 Design For Asia Bronze Award for the Dragon Skin Pavilion, and received the Perspective 2012 40 under 40 Award. LEAD is best known for their recently built "Golden Moon" Mid-Autumn Festival Lantern Wonderland 2012 in Victoria Park, Hong Kong. LEAD combines practice with research by organising and teaching in an International Workshop Series (IWS) with now 18 editions completed worldwide in countries as varied as Chile, Finland, South Africa and China.

LIKEarchitects

LIKEarchitects® is an award-winning practice focused on the design of ephemeral architectural objects and on socialy relevant international competitions.

The collective of architects, of an experimental, provocative and innovative nature, is formed by the young Portuguese architects Diogo Aguiar, João Jesus and Teresa Otto and seeks to combine their basilar architectural knowledge acquired in the renowned Faculty of Architecture of Oporto with other more radical architectural experiences they have had in worldwide reference studios such as UNStudio and OMA in The Netherlands and RCR Arquitectes, in Spain.

The proposed architecture, which are attentive to the current socio-economic scenario, aim to boost places and involve the community in a critical participation of urban space, having Installation, Happening and Urban Art as references.

LIKEarchitects' work has been awarded several prizes and been published both in national and international specialized magazines and books.

Los Carpinteros

The Havana-based collective Los Carpinteros (The Carpenters) has created some of the most important work to emerge from Cuba in the past decade. Formed in 1991, Los Carpinteros (consisting of Marco Castillo, Dagoberto Rodríguez, and, until his departure in June 2003, Alexandre Arrechea) adopted their name in 1994, deciding to renounce the notion of individual authorship and refer back to an older guild tradition of artisans and skilled laborers. Interested in the intersection between art and society, the group merges architecture, design, and sculpture in unexpected and often humorous ways. They create installations and drawings that negotiate the space between the functional and the non-functional. The group's elegant and mordantly humorous sculptures, drawings, and installations draw their inspiration from the physical world—particularly that of furniture. Their carefully crafted works use humor to exploit a visual syntax that sets up contradictions among object and function as well as practicality and uselessness.

Los Carpinteros' works are part of the permanent collections of the Solomon R. Guggenheim Museum, New York; The Museum of Modern Art, New York; The Los Angeles County Museum of Art; The Museum of Contemporary Art, Los Angeles and The Tate Modern, London, among others. Free Basket, a site-specific work commissioned by the Indianapolis Museum of Art in 2010, is permanently installed in the 100 Acres Park on the museum's grounds. They are the subject of a major monograph, Los Carpinteros: Handwork Constructing the World, published by Thyssen-Bornemisza Art Contemporary and Walther König in 2011.

Manuelle Gautrand

Manuelle Gautrand was born on July 14, 1961 in Marseille (France). She obtained her graduate diploma in Architecture from the "Ecole Nationale Supérieure d'Architecture de Montpellier" in 1985. She worked for 6 year in different architecture studios in Paris.

She founded her office in 1991, first in Lyons and then in Paris. She lives and works in Paris since 1994. She is the principal architect and director of the agency MANUELLE GAUTRAND ARCHITECTURE. She mainly designs buildings in areas as diverse as cultural facilities (theaters, museums, and cultural centers), office buildings, housing, commercial and leisure facilities, etc...

Her clients are public contracting authorities as well as private firms, in France and abroad.

In 2007 Manuelle Gautrand's the "C42" Citroen Flagship Showroom on the Champs-Elysées Avenue in Paris gained attention and widespread acclaim in the international arena and from a large audience.

MAS Arquitectura

Since 2000, the MAS Study architecture provides an integrated set of services, architectural design, interior design and construction, with the commitment to provide efficient and quality service, adapting their activities to the needs of our customers.

• 12 professionals with experience in different fields of architecture, design and construction

• More than 300 individuals and companies have relied on us

• More than 250 projects in over 50 different municipalities throughout Spain

Maya Nishikori + Yushiro Arimoto

Nishikori was born 1976 in Ehime, graduated from Tohoku University Biology 1999, Architecture 2001, completed master course in Tokyo University of the Arts 2002, Toyo Ito & Associates Architects 2002-2008, established own office 2009, Research Assistant of Tohoku University 2010.

Arimoto was born 1976 in Hyogo, graduated from Tohoku University Architecture 2000, completed master course of Urban & Architecture 2002, Coelacanth and Associates (C+A) 2003-2006, established own office 2008.

Both of them transferred to architecture from different fields. Nishikori had majored biology and Arimoto applied Physics. Their drafting boards were laid next to each other because they were transfer students. Arimoto studied medical planning and design and Nishikori did biology, architecture and art. They started to work together in 2010 for G Clinic. Now they are preparing an associate office. G Clinic 7f is awarded JCD 2012 best 100.

MET Studio

MET Studio celebrates its 30th year in business in 2012. It was originally set up in 1982 by its Chairman and founder Alex McCuaig, taking its name from the Metropolitan Wharf in London where the company's first design studio was sited. The company enjoys a global reputation, with a lengthy list of highly-acclaimed masterplanning, museum, exhibition, visitor centre, zoo, special event, retail, AV and branding projects, created for corporate clients such as Cunard, De Beer, Virgin, Lucent, Swire, Hongkong Telecom, Portugal Telecom and The Wellcome Trust, as well as non-corporate clients including The Dutch, Irish and Macau Governments, the local councils of Hull and Birmingham and the London Borough of Southwark, as well many national museums around the world. MET Studio has won or been a finalist in 65 individual awards and a winner of over 30 awards since the company began, including some of the world's most prestigious business and design awards, such as The Queen's Award for Enterprise and the Business Link Exporter of the Year Award, as well as being a winner at The Museum of the Year Awards, The DBA Design Effectiveness Awards (which measure the tangible results of design) and taking Design Week's top plaudit — the Best of Show Award — for the best design of any kind in a 12-month period.

The company has offices in London and Hong Kong.

Minas Kosmidis

Studio Minas Kosmidis [Architecture In Concept] is the evolution of the architect's private studio and its relocation, in 2007, from Komotini to the city of Thessaloniki.

Architect Minas Kosmidis graduated from the School of Architecture of the Aristotle University of Thessaloniki in 1988 and in 1991 he completed his postgraduate studies in "Industrial Design", with a scholarship from "EOMMEX", at the École d' Architecture de Paris-Conflans in the department of "Étude et Creation de Mobilier". During his Master studies he worked together with architect & designer Galeriste Lamia Hassanaein, in Cairo and Paris.

In 1993 he establishes his architectural studio in the town of Komotini, with operations around Greece. The studio undertakes private projects, both housing and professional, with a specialization in the area of dining and entertainment establishments.

Using the abstraction, the neatness of lines, the clarity, the transparency, the symmetry, the flow, the balance of volumes, the elements of nature and the light as tools, he is inspired to create unique projects, which combine full functionality and unique aesthetics.

The studio's projects have been published in many architecture, interior-design magazines and websites in Greece and abroad.

NAU | Architecture & Design

NAU is an international, multidisciplinary design firm, spanning the spectrum from architecture and interior design to exhibitions and interactive interfaces. As futurists creating both visual design and constructed projects, NAU melds the precision of experienced builders with the imagination and attention to detail required to create innovative exhibits, public events and architecture.

NAU has quickly garnered recognition as an accomplished creator of fashionable interiors for retail, hotels, restaurants and residences. Its dedicated teams offer a personal touch, working with clients to align design approach with the appropriate market. Distilled in clear, contemporary forms, the designs of NAU promote modern, flexible solutions that engage and welcome.

Outofstock Design

Outofstock is an international collective of award-winning designers based in Singapore, born out of a meeting in Stockholm — hence the name Outofstock.

Gabriel Tan and Wendy Chua from Singapore, Gustavo Maggio from Argentina and Sebastián Alberdi from Spain met at Electrolux Design Lab, Stockholm, in 2005. What started out as a cross cultural creative experiment grew into a design studio with offices in Singapore and Barcelona. They currently work on furniture, industrial design and interior design for international clients such as Ligne Roset, Environment, Quantum Glass, and Discipline.

Their accolades include Elle Decoration Spain's Young Designer of the Year 2008, the Singapore Furniture Design Award 2009 open category first prize and the President's Design Award 2010, Design of The Year.

P-06 Atelier

P-06 Atelier is an international award-winning firm specializing in communication and environmental design on a wide range of scales. Based in Lisbon, Portugal, the studio was founded in 2006 by partners Nuno Gusmão, Estela Estanislau, Pedro Anjos and Catarina Carreira. It has since undertaken a variety of projects from complex, large scale way finding systems, museum and exhibition design, to communication and editorial design for the printed page, with a bold, striking style that has garnered a number of distinctions. P-06 Atelier actively engages in collaborations with architects, urbanists, landscape designers and engineers, in a continuous, seamless workflow with complementing disciplines, enriching the firm's scope of work and amplifying every intervention's outcomes.

Pierluigi Piu

Pierluigi Piu was born in Cagliari (Sardinia, Italy) in 1954 and pursued his studies at the University of Architecture in Florence. He opened his own office and began working in the field of interior design and architecture in 1991. Then, from 1996 until 1998, he was back in Bruxelles, where he had been summoned by the architect Steven Beckers to collaborate on a project for the reconstruction and refurbishment of the Berlaymont Palace, the historic seat of the Council of Ministers of the European Community, and so undertook the supervision and coordination of the aesthetic and formal language for the interior design of the entire building.

He has won many awards, including the "Russian International Architectural Award 2007" in Moscow, the "International Design Award 2008" in Los Angeles, the "Archi-Bau Design Award 2009" in Munich, Germany, the "Compasso d'Oro 2011" in Rome and the "Premio IED" award from the international school Istituto Europeo di Design in 2012.

Reiulf Ramstad Arkitekter

Reiulf Ramstad Arkitekter (RRA) is an independent architectural firm with a high level of expert knowledge and a distinct ideology. RRA plays a leading role (in Norway) within its field. The office has a strong conceptual approach combined with experience from past accomplished projects. Over the years the office has produced a wide range of innovative and ground breaking projects and has experienced the overall process, from the first concept phase to the completion of high quality projects. The firm has shown a multitude of approaches in solving assignments, both national and international and has received numerous of prizes and awards for their projects.

Rosan Bosch

As a Dutch-born artist, Rosan Bosch has had an international launch pad for her professional work. She is partly educated at Hogeschool voor de Kunsten, Utrecht, Holland, and partly at Universitat de Bellas Artes, Barcelona, Spain. Before settling in Denmark, Rosan Bosch has among other places lived in Spain and Belgium for many years. Rosan Bosch now lives in Copenhagen, Denmark, where she has been partner in the art and design company Bosch & Fjord from 2001- 2010. The 1st of January 2011 she established Rosan Bosch Studio.

Salvetti Studio

Andrea creates works of art that reflect the themes of nature and the environment using a number of materials and varied techniques paying particular attention to fusion metals. All this from his sculpture and design studio, renovated from an old jute factory outside Lucca, working closely with young assistants and his wife Patrizia.

He finds his main inspiration from the hills and woods and the links with the environment where he lives and this puts its mark on the themes and aims of his projects. He is involved at all stages of the creative process, from the initial idea to the realisation of the piece in the workshop, including the descriptions and comments, upholding the tenet that "know-how" guarantees awareness and independence of one's work.

Andrea has been a proponent of "self-production" for some time now and this has allowed him to explore a wide interdisciplinary space. His pieces can be labelled as sculpture and design, touching architecture and cooking, thus creating horizontal links between these varied fields.

Slade Architecture

Slade Architecture was founded in 2002, seeking to focus on architecture and design across different scales and program types. Their design approach is unique for each project but framed by a continued exploration of primary architectural concerns.

As architects and designers, they operate with intrinsic architectural interests: the relationship between the body and space, movement, scale, time, perception, materiality and its intersection with form. These form the basis of our continued architectural exploration.

Layered on this foundation, is an inventive investigation of the specific project context. Our broad definition of the project context considers any conditions affecting a specific project: program, sustainability, budget, operation, culture, site, technology, image/branding, etc.

Working at the intersection of these considerations, they create designs that are simultaneously functional and innovative.

They have completed a diverse range of international and domestic projects and their work has been recognized internationally with over 200 publications, exhibits and awards. The awards, publications and exhibits are a welcome recognition. However, they strive to create work that speaks for itself.

Softroom Architects

Softroom Architects have developed a reputation for design excellence, creative innovation and skilful delivery. Recent projects have seen the practice working on large-scale and international schemes of increasing size and complexity.

Founded in 1995, the company is led by two directors, Christopher Bagot and Oliver Salway, backed by an enthusiastic and committed team. Work by the practice has been published and exhibited worldwide.

Softroom have worked on a wide spectrum of projects including buildings, bridges and interiors for public, educational, residential, retail and commercial uses, transport and exhibition design. Working with a broad range of brands, public and private clients have resulted in a portfolio that is regarded as both stylish and imaginative.

Softroom have received many accolades during their career, including RIBA awards, the Stephen Lawrence Prize, "Building of the Year" from the Royal Fine Arts Commission, "Best in Show" from Design Week and prestigious "Yellow Pencils" from D&AD.

Spatial Ops

Spatial Ops is an intermittent collaborative between Steven Christensen Architecture and Anya Sirota / AKOAKI. Informed by the team members' diverse backgrounds in architecture, photography, psychology, and film, their work explores the relationship between space, media, and perception. The team is based in Los Angeles and Ann Arbor, and includes faculty from the UCLA and the University of Michigan. Their tactical design interventions seek to catalyze conversations about the relationship between urbanism, individual liberty, and contemporary culture.

SPEECH Tchoban & Kuznetsov

The bureau SPEECH Tchoban & Kuznetsov was founded in 2006 as a result of long term collaboration of directed by Sergei Tchoban Berlin office nps tchoban voss and his Moscow office "Tchoban and partners" with the bureau "S.P.Project", headed by Sergey Kuznetsov.

SPEECH Tchoban & Kuznetsov is one of the leading architectural bureaus in Russia. It specializes in designing of buildings and complexes of different functional purposes, in development of urban planning concepts and in creating interior decisions as well. Success in realization of projects, working up by the bureau, is contributed by great experience of work of Sergei Tchoban and Sergey Kuznetsov in Germany and Russia that allows adapting working knowledge of modern western materials and building technologies to the specific of project realization in Russian conditions. An important part depends on rational organization of designing process, on employees' expertise and technological level of the office.

At present staff of the bureau numbers nearly 200 employees — architects, constructors, engineers and managers developing tens of projects for Moscow, Saint-Petersburg and other cities of Russia, CIS and other parts of the world. Joining of different specialists in one team gives the bureau a possibility to develop complex decisions and participate in all stages of work at the project. As general designer in urban planning projects bureau SPEECH Tchoban & Kuznetsov gets involved in the work of the best Russian and foreign architects, constructors, engineers and specialists from other related sectors.

Collaboration with the bureau guarantees highly qualified planning of projects with different complexity in accordance with international standards, optimization of their execution periods, application of modern western technologies, strict authorial control of realization process of the projects.

Stamberg Aferiat + Associates

Founded in 1989 by Peter Stamberg and Paul Aferiat, Stamberg Aferiat + Associates is a comprehensive design firm based in New York City. Stamberg and Aferiat defy the predictable by synthesizing form, light, color and context into clear and compelling architectural experiences. For three decades they have received widespread acclaim for projects that captivate global audiences and challenge the status quo in residential, hotel, commercial and cultural typologies.

While still remaining at the forefront of design conversation, Stamberg Aferiat's work is be spoken to clients' aspirations and realities. The result is an incomparable group of built works ranging from the cubist forms at the Shelter Island Pavilion to an elegant and minimal phased expansion to one of Richard Meier's earliest works, the Hoffman House.

The Stamberg Aferiat team works across a variety of disciplines that include architecture, interiors, furniture, exhibition, textiles and graphic design. Their 1994 Salsa Sofa Collection is produced through Knoll. An updated edition of their award-winning monograph published by Rizzolli features essays by art and architecture icons Paul Goldberger, Joseph Rosa, David Hockney, Richard Meier and Charles Gwathmey. Their hospitality work include the Saguaro Hotels where they developed and implemented the brand across multiple scales.

Stone Designs

Stone Designs is the story of two creators (Cutu Mazuelos 1973 and Eva Prego 1974) who in September 1995 started their adventure by creating their own studio, in order to develop their projects from a personal perspective, without censureship or interference. For this reason they never worked for anyone else and they forged their apprenticeship as it was done in the past, from below and with a lot of sacrifice. They began by doing interior designs, stands and displays. This field served them during the first years of their career, experimenting and studying different languages, showing a new understanding of design and our relationship with it.

This allows them to concentrate exclusively on what they do best, designing. It has left them time to take on projects of much greater depth and to work with companies with which they truly feel an affinity with. Being independent has its advantages... From companies of friends that they love and admire for their dedication to designing premises for companies such as Lexus, Telepizza, Havaianas, Coca Cola, and furniture for brands such as the Japanese Muji, the Swedish Offecct of the Spanish RS.

STUDIO 38

The core know-how is in the area of retail and corporate design, as well as visual merchandising and the creation of brand staging. 2D and 3D developments are in demand just as much as individual IT solutions. From order handling through product support up to delivery of the requested product, studio 38 takes on the complete monitoring, also for Europe-wide and worldwide rollouts.

The qualified communication designers Yannah Bandilla, Kathrin Janke-Bendow and Reinhard Knobelspies have been working together for over 15 years. In 1996 they combined their different strengths, whether of creative, strategic or analytical origins and founded studio 38.

The studio 38 team is made up of communication, product and fashion designers, interior designers, account managers and project managers as well as programmers, who are teamed up individually together for each project based on their interdisciplinary orientation, in order to continuously achieve successful and innovative communication solutions.

Synthesis Design + Architecture

Synthesis is an emerging contemporary design practice with over 20 years of collective professional experience in the fields of architecture, infrastructure, interiors, installations, exhibitions, furniture, and product design. Founded in 2011, our work has already begun to achieve international recognition for its design excellence. Our diverse team of multidisciplinary design professionals includes registered architects, architectural designers and computational specialists educated, trained, and raised in the USA, UK, Denmark, Portugal, and Taiwan. This diverse cultural and disciplinary background has supported our expanding portfolio of international projects in the USA, UK, Russia, Thailand, and China. Since its founding, the office has expanded to include a team of 7 full-time designers with a diverse global background and wide range of experiences. The design trajectory and ethos of the office is rooted in balancing both the experimental and the visionary with the practical and the pragmatic to achieve the extraordinary.

UXUS Design

Established in 2003, UXUS is a leading global strategic design consultancy delivering innovative consumer experiences for top multi-national brands. Regarded internationally as a thought leader, UXUS produces emotional and intelligent design exemplifying the principles of Brand Poetry: balancing creative excellence with commercial success.

The award-winning team at UXUS consistently delivers the most unique and exciting solutions possible, creating noteworthy design and breaking through industry standards. UXUS has built a reputation for design excellence and innovation, combined with a comprehensive perspective into the world of consumer experiences and products.

UXUS works with cutting-edge global brands in over 40 countries, including Qatar Luxury Group, Selfridges, Bloomingdale's, Chanel, InterContinental Hotels Group, Tate Modern, P&G, McDonald's and many other valued clients.

In 2013, UXUS joined partnership with FutureBrand, a leading brand consultancy network with over 25 offices worldwide.

Zhang Tao

Zhang Tao has served in domestic and overseas firms of architectural design and construction successively, such as Haigo Shen International Engineering Consultants Inc., Shimaogroup, K. F. Stone Design International Inc. Canada, and so on. He has taken charge of Shanghai Science & Technology Museum, Shimao Lakeside Garden, Shimao Village Garden, Shell China Exploration and Production Company Limited (Shell), Millennium Hotel Wuxi and (other interior and exterior) design projects.

Zhang Tao is skilled in creating a comfortable natural human space by using various techniques. With cleverly taking advantage of environment, color and materials, he can make the color and lighting effects present different charms in space. In addition, designer also can masterly combine design elements with nature among the space, which gives people with the modern, fashionable, concise, graceful feeling. It is his professional spirit to the integrity and unity of design that makes his projects have won high praise from both owners and the industry.

THERE Design

THERE

THERE is a design agency specializing in creating and transforming brands, developing assets and experiences that enable a better connection with their audience.

Driven by strategic and conceptual insights, our integrated approach delivers solutions that provoke thought and inspire action, enabling our clients to turn their brand into one of its greatest assets, seeing real value and ROI.

What we do may be common, how we do it, is not. We challenge the ordinary with our combination of enthusiasm, intelligence, creativity and range of execution.

We've been doing this since 2002 and continue to deliver progressive and multi-award winning outcomes for a diverse client list that includes, ASICS, 3M, SAP, Wilson HTM Investments, Ontera Modular Carpets, McGrath Real Estate, McDonalds, Colliers, Cochlear, Orient Express Hotels, Macquarie Bank and Carnival Group to name a few.

Services we offer include: brand strategy and positioning, brand identity, brand communications, website and digital media, branded environments, workplace graphics, wayfinding and signage.

Wesley Meuris

In his work Wesley Meuris starts out from the interaction between architecture and our conditioned behavior. He questions current conventions and the automatic approach we take to standardized architectural spaces. His starting point is the basic rules that have taken shape in the course of time regarding the dimensions, materials, proportions and division of the environment we live in. He plays with these conventions (cultural and otherwise) and questions them. In this way he sees to it that familiar forms and buildings come across as both recognizable and strange.

ARTPOWER

Acknowledgements

We would like to thank all the designers and companies who made significant contributions to the compilation of this book. Without them, this project would not have been possible. We would also like to thank many others whose names did not appear on the credits, but made specific input and support for the project from beginning to end.

Future Editions

If you would like to contribute to the next edition of Artpower, please email us your details to: artpower@artpower.com.cn